Oxygen Free Radicals in Shock

International Workshop on Oxygen Free Radicals in Shock,
Florence, May 31–June 1, 1985

Oxygen Free Radicals in Shock

Editors
G. P. Novelli, Florence
F. Ursini, Padova

63 figures and 48 tables, 1986

 KARGER

Basel · München · Paris · London · New York · New Delhi · Singapore · Tokyo · Sydney

National Library of Medicine, Cataloging in Publication
International Workshop on Oxygen Free Radicals in Shock (1985: Florence, Italy)
Oxygen free radicals in shock / International Workshop on Oxygen Free Radicals in Shock,
Florence, May 31–June 1, 1985; editors, G. P. Novelli, F. Ursini. – Basel; New York: Karger,
1986.
Includes index.
1. Free Radicals – congresses 2. Oxygen – poisoning – congresses 2. Shock – metabolism –
congresses I. Novelli, Gian Paolo II. Ursini, F. III. Title
QV 312 I615o 1985
ISBN 3–8055–4233–X

Drug Dosage
The authors and publisher have exerted every effort to ensure that drug selection and dosage
set forth in this text are in accord with current recommendations and practice at the time of
publication. However, in view of ongoing research, changes in government regulations, and
the constant flow of information relating to drug therapy and drug reactions, the reader is
urged to check the package insert for each drug for any change in indications and dosage and
for added warnings and precautions. This is particularly important when the recommended
agent is a new and/or infrequently employed drug.

Contents

Contents

Contents

Foreword

The contents of the present volume include the formal presentations of the oral communications and posters delivered at the workshop on Oxygen Free Radicals in Shock, held in Florence May 31st to June 1st, 1985. This Workshop was the first in what we hope will be a long series of scientifically valuable meetings arranged under the auspices of the newly formed European Shock Society.

Prof. *Gian Paolo Novelli* and his co-workers are to be congratulated for putting together an admirable meeting and for helping our Society in their initial venture. He has set a standard which all of the succeeding organizers can well emulate.

As the current President of the European Shock Society it was my duty and my privilege to attend this Workshop. It was both scientifically and socially an enormous success. Most importantly it pointed the way to new avenues of approach in the investigations of the pathophysiology and treatment of shock. We all benefited from the meeting and will also benefit from having these Proceedings in our hands.

David H. Lewis, MD
President, European Shock Society

Preface and Acknowledgements

In September 1984 – during the Manchester Meeting of the European Shock Society – the idea of a Workshop on Oxygen Free Radicals in Shock began to grow up. In May 1985, due to the enthusiastic adhesion of many scientists in the whole world, the workshop was held in Florence during three beautiful days. The site of the meeting (offered by the Accademia 'La Colombaria', founded in 1735) was built in the XIIIth century as one of the first Italian hospitals and the conference hall was an hospital ward.

I am grateful to Accademia and to Prof. *Francesco Adorno* for giving the authorization to make use of such an exceptional old site. I am grateful also to the University of Florence and to Rector Prof. *Franco Scaramuzzi* for the support to this workshop. I thank also Mrs. *Cinzia Giachè* and Mrs. *Patrizia Pellegrini,* secretaries, for their enthusiastic, continuous cooperation.

The organization was made possible by the financial contributions of many institutions – included the National Research Council (CNR) and the Tuscany Region – and companies.

But the main aim of this brief speech is to remind all the readers of my wife, *Sofia,* whose clear suggestions and criticism were primary for the final result of this meeting. She died one month before the workshop and now I dedicate this book to *Sofia.*

Florence, 31st May 1985   *Gian Paolo Novelli*

Sponsors of the Conference

Novelli, Ursini (eds.), Oxygen Free Radicals in Shock. Int. Workshop,
Florence 1985, pp. 1–8 (Karger, Basel 1986)

Biochemical Mechanisms of Oxy-Radical Production and the Role of the Antioxidant Enzymes in Relation to Hypoxic and Ischemic Tissue Damage

Giuseppe Rotilio

Department of Biology, II University of Rome, 'Tor Vergata', Rome, Italy

Biological Activation of Molecular Oxygen

The formation of oxy-radicals in the human body is a consequence of aerobic life. This is, in turn, an effect of the accumulation of oxygen in the terrestrial atmosphere after the proliferation of photosynthetic organisms, which reached a significant extent some 500 million years ago. Circulatory systems with oxygen-carrying molecules such as hemoglobin, and respiratory systems were developed in metazoa just at this age. Presently, because of its ubiquitous availability and favorable thermodynamic potential, atmospheric oxygen is utilized by most multicellular life forms as the ultimate acceptor of electrons from the reduced carbon chains of carbohydrates and lipids, in a process occurring inside specialized membranes and leading to adenosine triphosphate (ATP) synthesis. H_2O is the product of oxygen reduction in this process, which involves the transfer of 4 electrons to the molecular oxygen or dioxygen (O_2). It is well known that this reaction is kinetically unfavorable, and requires activation of oxygen by specialized enzyme systems containing metal ions as prosthetic groups. According to the quantum mechanical rules of spin restriction the reduction of dioxygen by one electron at each time (univalent pathway) is much more facile [1]. However, it leads to the formation of very reactive intermediates (fig. 1), which will not be compatible with life, if formed in high concentration.

The enzymes of oxygen activation in biological membranes (e.g. cytochrome oxidase, in mitochondria, cytochrome P-450, in endoplasmic

Dioxygen	Superoxide	Hydrogen peroxide	Hydroxyl radical	Water

$$O-O \xrightarrow[+\ 1H^+]{+\ 1e^-} HO-O^{\cdot} \xrightarrow[+\ 1H^+]{+\ 1e^-} HO-O-H \xrightarrow{+\ 1e^-} OH^{\cdot} \xrightarrow{+\ 1e^-} H-O-H$$
$$+$$
$$OH^-$$

$(O_2^-$ at
physiological
pH values:
$O_2^- \rightleftarrows HO_2^-$
pK = 4.8)

Fig. 1. The univalent pathway of oxygen reduction.

reticulum) are able to trap the reactive intermediates, superoxide, hydrogen peroxide and hydroxyl radical, inside their active site pockets, until all of the four electrons required for water production are transfered to dioxygen. This 'coupling' is not always perfect, however, and liberation of intermediates of the univalent reduction has been shown to occur in cells and tissues under several conditions.

Potential Sources of Oxy-Radicals in the Body (table I)

Physical Sources. When the body is exposed to sources of electromagnetic radiations, such as X-ray or light (in the presence of photosensitizers, such as porphyrins in porphyrias), oxy-radicals may form according to two mechanisms. A mechanism involves a direct radiolysis of the body water ($H_2O \xrightarrow{h\nu} H^{\cdot} + OH^{\cdot}$) and further reactions of $H^{\cdot}$ with O_2^- to give O_2^- This mechanism requires very high energy, such as the X-ray doses used in cancer therapy. A more frequent case involves the interaction of organic molecules present in the body with light or radiations. This interaction either leads to excited states, which are able to transfer an electron to O_2 with formation of O_2^-, or actual photolysis of H atoms from the molecule, with the formation on an organic free radical that in turn reacts with O_2 giving O_2^-.

Chemical Sources. There are two distinct cases of chemical production of oxy-radicals: inorganic ions and organic molecules. In both cases, the chemical species must be able to undergo a univalent redox cycle. Inorganic couples of this kind are Cu(I) – Cu(II) or Fe(II) – Fe(III), which are present in the body not only as prosthetic groups of metalloenzymes

Table I. Potential sources of oxy-radicals in living organisms

1 Physical agents

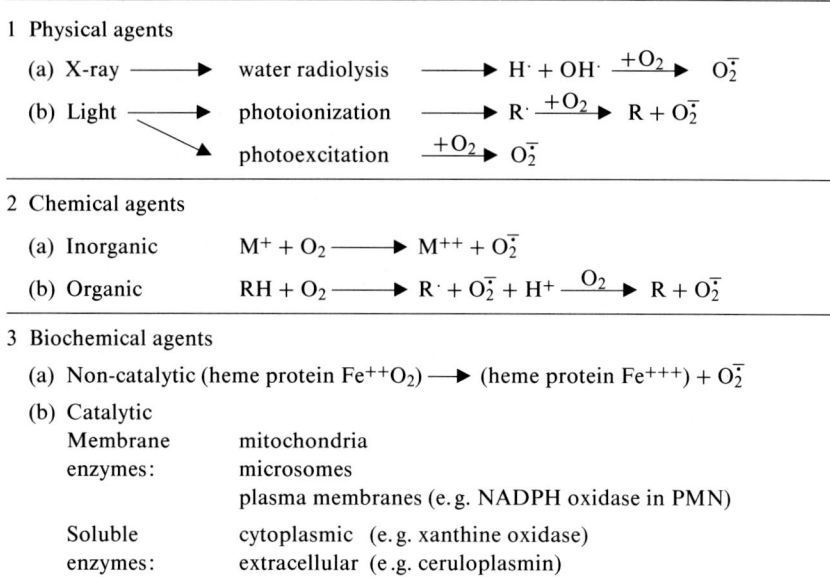

2 Chemical agents

(a) Inorganic $\quad M^+ + O_2 \longrightarrow M^{++} + O_2^-$

(b) Organic $\quad RH + O_2 \longrightarrow R^{\cdot} + O_2^- + H^+ \xrightarrow{O_2} R + O_2^-$

3 Biochemical agents

(a) Non-catalytic (heme protein $Fe^{++}O_2$) $\longrightarrow$ (heme protein Fe^{+++}) $+ O_2^-$

(b) Catalytic

Membrane enzymes:	mitochondria
	microsomes
	plasma membranes (e.g. NADPH oxidase in PMN)
Soluble enzymes:	cytoplasmic (e.g. xanthine oxidase)
	extracellular (e.g. ceruloplasmin)

but also as low molecular weight complexes. Organic univalent redox couples include biological quinones (e.g. CoQ, Vit K), thiols (e.g. glutathione), flavins, ascorbic acid. The low valence states of such couples can react with oxygen, and O_2^- will be generated in this process.

Biochemical Sources. Biochemical systems can generate oxy-radicals either stoichiometrically or catalytically. *Non-catalytic* sources are mainly the oxygenated forms of heme proteins [2], and in particular oxyhemoglobin [3]. The oxygen adducts of heme proteins are formally superoxo-Fe(III) complexes, i.e. $Fe(II)-O_2 = Fe(III)-O_2^-$. O_2^- may dissociate from the complex in the presence of high concentrations of nucleophiles (e.g. Cl^-), or may be produced by redox reactions of HbO_2 with Cu(II) or drugs (e.g. primaquine). *Catalytic* biochemical sources of oxy-radicals are the numerous electron-transferring enzymes which are able to reduce oxygen. Two distinct situations are to be considered in this respect. A class of redox enzymes seem to be competent for the univalent reduction of oxygen in normal conditions. A typical example is the NADPH oxidase of the plasma membrane of neutrophils, which is elicited by phagocytic stimuli to produce large yields of O_2^- as a biological weapon against aggressive agents [4]. The second group is much larger and includes systems that switch to O_2^- production under abnormal conditions. This is the case of the mitochondrial chain in the presence of uncouplers [5], of the microsomal hydroxylating chain in the presence of redox-cycling drugs such as adriamycin and daunomycin [6], of the cytoplasmic xanthine dehydrogenase when transformed to xanthine oxidase in ischemic tissues [7].

Reactions of Oxy-Radicals in Living Organisms

O_2^- has been shown to peroxidize lipids, depolymerize polysaccharides, inactivate enzymes, and break DNA. Hemolysis, bacterial killing and death of mammalian cells are established effects of exposure of living systems to O_2^- [8]. The chemical basis of O_2^- toxicity is its extremely efficient reactivity in one-electron exchange reactions, the most important of which are the following:

$$O_2^- \text{ dismutation}: O_2^- + O_2^- \xrightarrow{2H^+} H_2O_2 + O_2.$$

$$\text{Metal reduction}: O_2^- + M^{++} \longrightarrow O_2 + M^+.$$

Therefore, the production of O_2^- in biological systems, where metal ions are ubiquitous and protons easily available, leads to immediate production of H_2O_2 and reduced metal ions. These in turn react with each other, according to the well known Fenton reaction: $H_2O_2 + M^+ \rightarrow OH^{\cdot} + OH^- + M^{++}$.

OH$^{\cdot}$ is considered to be the actual damaging species in systems where fluxes of O_2^- are produced, because O_2^- and H_2O_2 per se are unable to degrade biological molecules. On the other hand, the feasibility of the reaction described above in biological systems has been demonstrated [8], and presently these reactions are believed to be responsible for oxygen cytotoxicity and oxy-radical damage.

The Enzymes that Combat Oxy-Radical Injury in Living Organisms

The evolution process that led to development of photosynthetic organisms and enrichment of terrestrial atmosphere in O_2 was possible because of specialized metalloenzymes capable of oxidizing H_2O with liberation of O_2. Metalloenzymes were also utilized to cope with the increased risk of oxy-radical fluxes, which faced aerobic cells since the beginning of the aerobic age. The essential enzymatic defense in mammalian cells is presently identified with two catalytic systems [9]:

Superoxide Dismutase (SOD): $O_2^- + O_2^- \xrightarrow{2H^+} O_2 + H_2O_2$. This reaction is catalyzed by an enzyme carrying Cu and Zn on each of two equivalent subunits of 16,000 daltons and present in all cell compartments but plasma membranes and peroxisomes. It is able to catalyze superoxide dismutation with much greater efficiency than the proton-catalyzed 'spontaneous' reaction. This means that O_2^- is removed by the enzyme even at pH $\geq$ 7 and at very low [O_2^-] concentration. These points are of high physiological relevance (see below). Another superoxide dismutase contains Mn and is present in the mitochondria of most mammals, but in primates and man appears to be evenly distributed in mitochondria and cytoplasm. It is an important enzyme because, although much less efficient than the Cu-Zn enzyme, it is inducible by high concentrations of oxygen [8].

Glutathione Peroxidase (GSH Px): $2 \text{ GSH} + H_2O_2 \rightarrow \text{GSSG} + 2H_2O$. This enzyme [10] is responsible for the removal of H_2O_2 in all compartments but peroxisomes. It is also able to transform organic peroxides into the corresponding alcohols and is therefore the major protective enzyme against lipid peroxidation. It contains Se as inorganic cofactor, in the form of a selenocysteine residue at the active site. Its function is supported by other enzymes, which are needed to regenerate GSH: GSSG reductase, an NADPH-dependent enzyme, and the NADPH-producing dehydrogenases.

It is very likely that SOD and GSH Px are the components of a single functional system, aimed to minimize O_2^- and H_2O_2 concentrations in tissues and thus limiting the risk of OH$\cdot$ production. It should also be kept in mind that both O_2^- and GSH are efficient producers of reduced metal ions and organic free radicals, both potential sources of OH$\cdot$ radicals. Thus, the two enzymes also help maintaining metal ions and redox cycling organic substrates and cofactors in the less dangerous oxidized state. The functional linkage of the two enzymes has been recognized during mammalian development [11] and differentiation [12], and in several pathological models [13–15]. It has been shown that a relative increase of the SOD/GSH Px ratio leads to increased sensitivity of cells to O_2^- damage, probably because the enzymatically unbalanced cell cannot cope with higher fluxes of H_2O_2.

The Effects of Lack of Oxygen on Oxy-Radical Damage of Tissues

Tissue damage by oxy-radicals depends on the magnitude of oxy-radical fluxes, on the enzymatic capability of a certain tissue to face an increased flux of oxy-radicals and on other factors which are not constitutive of the system but may be defined as circumstantial or environmental components. Availability of trace metal ions in the form of low molecular weight chelates, able to undergo facile redox reactions with O_2^- and H_2O_2, is certainly among the additional factors and is of crucial importance to the production of $OH^\cdot$ radicals. In recent years, the ischemic state has been the object of a great deal of studies centered on oxy-radical damage. Such a paradoxical connection is, however, supported by several lines of evidence. I shall just list some possible mechanisms that may play a role to make oxy-radical damage possible, and even more likely, in conditions of relative lack of oxygen supply to tissues.

(a) When the oxygen tension in mitochondria is below the normal value, the intermediate semiquinone states of flavines and ubiquinone live longer. The same happens for reduced states of GSH and metal ions. Readmission of oxygen in the reperfusion phase may lead to excessive univalent reduction of oxygen, also because coupling of electron transfer to the multivalent pathway is less tight after ischemic damage of membranes.

(b) Lack of ATP synthesis leads to increased concentration of precursors such as xanthine and hypoxanthine [7]. For the same reason the Ca^{++} pumps collapse and a redistribution of Ca^{++} in the cell takes place. This leads in turn to activation of cytosolic proteases. Among the protease effects, the conversion of xanthine dehydrogenase to xanthine oxidase has been described in several tissues [7]. In the reperfusion step, enhanced oxidation of xanthine produces very high fluxes of O_2^- in the post-ischemic tissue, which might overwhelm its enzymatic defense capabilities.

(c) During the ischemic impairment of blood flow, an augmented adhesion of granulocytes to blood vessels has been observed [16]. In the reperfusion step, the NADPH oxidase of their membranes is a potential source of an excessive flux of O_2^- in the microvasculature.

(d) At low O_2^- concentrations, as expected in the ischemic phase, 'spontaneous' dismutation is not very rapid, being a bimolecular process depending on $[O_2^-]^2$. In this condition O_2^-, although at low concentration, lives longer, and may trigger undesired reactions, such as protease activation [17], which, being a 'cascade' process, needs small quantities of triggering agents. If SOD is present, however, even minimal O_2^- fluxes can be stopped, due to the high efficiency of the enzyme even at very low substrate concentration. On the basis of these considerations the use of SOD as therapeutic agent is to be regarded favorably in both the ischemic and reperfusion steps of ischemic diseases.

References

1   Taube, H.: Mechanisms of oxidation with oxygen. J. gen. Physiol. *49:* 29 (1975).
2   Rotilio, G.; Falcioni, G. C.; Fioretti, E.; Brunori, M.: Decay of oxyperoxidase and oxygen radicals. Biochem. J. *145:* 405–407 (1975).
3   Brunori, M.; Falcioni, G. C.; Fioretti, E.; Giardina, B.; Rotilio, G.: Formation of superoxide in the autoxidation of the isolated α and β chains of human hemoglobin. Eur. J. Biochem. *53:* 99–104 (1975).
4   Babior, B. M.; Kipnes, R. S.; Curnutte, J. T.: Biological defense mechanisms: the production by leukocytes of superoxide. J. clin. Invest. *52:* 741–744 (1973).
5   Forman, H. J.; Boveris, A.: Superoxide radical and hydrogen peroxide in mitochondria; in Pryor, Free radicals in biology, vol. V, pp. 65–90 (Academic Press, New York 1982).
6   Bachur, N. R.; Gordon, S. L.; Gee, M. V.: Anthracycline antibiotic augmentation of microsomal electron transport and free radical formation. Molec. Pharmacol. *13:* 901–910 (1977).
7   McCord, J. M.: Oxygen-derived free radicals in postischemic tissue injury. New Engl. J. Med. *312:* 159–163 (1985).
8   Fridovich, I.: Superoxide radical and superoxide dismutase; in Gilbert, Oxygen and living processes, pp. 250–272 (Springer, New York 1981).
9   Forman, H. J.; Fischer, A. B.: Antioxidant defenses; in Gilbert, Oxygen and living processes, pp. 235–249 (Springer, New York 1981).
10 Flohe, L.: Glutathione peroxidase brought into focus; in Pryor, Free radicals in biology, vol. V, pp. 223–254 (Academic Press, New York 1982).
11 Mavelli, I.; Autuori, F.; Dini, L.; Spinedi, A.; Ciriolo, M. R.; Rotilio, G.: Correlation between superoxide dismutase, glutathione peroxidase and catalase in isolated rat hepatocytes during fetal development. Biochem. biophys. Res. Commun. *102:* 911–916 (1981).
12 Mavelli, I.; Rigo, A.; Federico, R.; Ciriolo, M. R.; Rotilio, G.: Superoxide dismutase, glutathione peroxidase, and catalase in developing rat brain. Biochem. J. *204:* 535–540 (1982).
13 Bozzi, A.; Mavelli, I.; Mondovì, B.; Strom, R.; Rotilio, G.: Differential cytotoxicity of daunomycin in tumour cells is related to glutathione-dependent hydrogen peroxide metabolism. Biochem. J. *194:* 396–372 (1981).
14 Mavelli, I.; Ciriolo, M. R.; Rotilio, G.; Superoxide dismutase, glutathione peroxidase and catalase in oxidative hemolysis. Biochem. biophys. Res. Commun. *108:* 286–290 (1982).
15 Mavelli, I.; Ciriolo, M. R.; Rossi, L.; Meloni, T.; Forteleoni, G.; De Flora, A.; Be-

natti, U.; Morelli, A.; Rotilio, G.: Favism: a hemolytic disease associated with increased superoxide dismutase and decreased glutathione peroxidase activities in red blood cells. Eur. J. Biochem. *139:* 13–18 (1984).

16   Grögaard, B.; Gerdin, B.; Arfors, K. E.: Involvement of neutrophils in the cortical blood flow impairment after cerebral ischemia in the rat. Proc. 4th Int. Conf. Superoxide and Superoxide Dismutase, Rome 1985.

17   Baehner, R. L.; Boxer, L. A.; Ingraham, L. M.: Reduced oxygen by-products and white blood cells; in Pryor, Free radicals in biology, vol. V, pp. 91–113 (Academic Press, New York 1982).

Prof. G. Rotilio, MD, Department of Biology, II University of Rome,
'Tor Vergata', I-00173 Rome (Italy)

Novelli, Ursini (eds.), Oxygen Free Radicals in Shock. Int. Workshop,
Florence 1985, pp. 9–14 (Karger, Basel 1986)

The Multilevel System against
Lipid Peroxidation in Living Tissues

Fulvio Ursini

Institute of Biological Chemistry of the University of Padova, Italy

Lipid peroxidation is an oxygen-dependent deterioration of fats. The
interest in this process, initially restricted to food technologists, in the
past few years rapidly increased following the recognization that peroxidative damage of biomembranes is a key factor in a number of diseases
and in aging [1, 2]. Moreover, *McCord and Fridovich's* [3] discovery of the
superoxide dismutase stimulated a great deal of research on superoxide
as well as on other oxy-radical species. These studies carefully defined
the chemical reactivity of the oxy-radicals and the chemistry of lipid peroxidation [4, 5]. The molecular mechanisms of the initiation, propagation
and termination reactions are generally accepted, at least in simple chemical or biochemical systems. On the other hand, there is still a major limitation to the quantitative measurement of lipid peroxidation in vivo,
where the peroxidation products are not stable and rapidly metabolized.
The evaluation of the role of free radicals on lipid peroxidation in aging
and diseases indeed suffered this limitation and so was assessed mainly
on the basis of the protective effect of antioxidants [1, 2]. However, so-
phisticated analytic approaches, such as the measurement of volatile hydrocarbons in the breath [6], the EPR analysis of perfused spin trappers
[7], or the ultraweak chemiluminescence emission from exposed organs
[8], confirmed that lipid peroxidation occurs in vivo. The imbalance between oxidative reactions and protection mechanisms in the cells is therefore the basis of the aging process and of a number of diseases.

The cellular defences against oxygen toxicity and lipid peroxidation
appeared progressively during the evolution and constitute a complex
multilevel system. In the aerobic cells indeed the oxygen is activated under carefully controlled conditions [9], there is a fine control of the con-

centration of metals, active as catalysts of peroxidation [10], there are specific enzymes scavenging the reactive intermediates of oxygen reduction [11, 12] and finally there are molecules (antioxidants) that terminate an oxidative chain reacting with radicals [13]. Moreover, the first protection mechanism against uncontrolled oxidative reactions is related to the electronic nature of oxygen itself. The ground state of the oxygen molecule is paramagnetic, and so in order to get an oxidative reaction with the most common diamagnetic molecules a change of the spin in one of the two unpaired electrons of oxygen must occur, but this event is restricted [14]. This 'spin restriction', which dramatically reduces the chemical reactivity of oxygen with organic molecules, can be overcome when oxygen is 'activated'. The reaction with a single electron reducing paramagnetic molecule (metals) or the conversion to the exicted singlet state removes the spin restriction and increases the chemical reactivity. Both the physiological metabolism and the toxicity of oxygen require this 'activation'. The metabolic oxygen activation occurs at the active site of specific enzymes reducing oxygen by two electrons (aerobic dehydrogenases) or by four electrons (cytochrome oxidase). In both cases the release of reactive intermediates is prevented. The perfect reaction mechanism of these enzymes represents the first defence against oxygen toxicity. However, a significant part of the oxygen in aerobic organisms can be reduced by a single electron-producing superoxide. Two enzymes were recognized as catalyzing the single electron reduction of oxygen: the leukocyte NADPH oxidase [15] and the xanthine oxidase, produced by a limited proteolysis from the anaerobic xanthine dehydrogenase [16]. The toxicity of superoxide is generally agreed to be dependent on the reaction with hydrogen peroxide in the presence of catalytic amounts of iron (Haber-Weiss reaction). The reaction proceeds through the reduction of ferric iron by superoxide. The ferrous iron then reacts with hydrogen peroxide giving rise to a hydroxyl ion and to a hydroxyl radical in a Fenton type reaction [17]. The removal of superoxide is catalyzed by superoxide dismutases [11], a widely distributed family of enzymes containing a transition metal (Cu, Fe, or Mn) undergoing a catalytic cycle coupled to the disproportion of superoxide to oxygen and hydrogen peroxide. The eukariotic cells contain a Cu-Zn superoxide dismutase (SOD) in the cytosol and a Mn SOD in the mitochondria. The hydrogen peroxide is produced by aerobic dehydrogenases by SOD and by the spontaneous disproportion of superoxide, and it is removed by catalase and glutathione peroxidases. The catalase is present in virtually all the mammalian cells where it catalyzes the disproportion of hydrogen peroxide, or a peroxidase reaction, in the presence of a suitable H donor (ethanol, phenols, formic acid). The glutathione peroxidase is a tetrameric enzyme containing Se-cysteine at the active site [12]. While the specificity for thiol is restricted to glutathione, a number of different peroxidic substates can be reduced. The glutathione transferase B [19] catalyzes the same reaction and is generally referred to as 'non-Se-dependent glutathione peroxidase'. This activity has been observed to increase when the availability of Se for the synthesis of glutathione peroxidase is low. The supply of reduced glutathione is, indirectly, another component of the defence system against free radical damage. A decreased content of reduced glutathione caused by defects in the synthesis or in the reduction leads to a hemolitic crisis following oxidative stress [20]. Moreover, the experimental glutathione depletion induces cell damage related to an increased lipid peroxidation [21]. In spite of the high efficiency of the enzymatic systems described, some lipid hydroperoxides may be formed and from these free radicals can be generated by reductive activation or by molecule-induced homolysis of the O–O bond [23]. The rate of these secondary initiations increases rapidly in the presence of metal complexes, and it was postulated that the iron complexes catalyzed breakdown of hydroperoxides is the major mechanism of lipid peroxidation initiation in vivo [22]. At this level the cellular defence must depend on the capacity to reduce the hydroperoxide groups in the membranes. This activity was first attributed to glutathione peroxidase, but it was recently ascertained that this enzyme is not active on phospholipid hydroperoxides [23] and cannot prevent secondary initiations in the membranes. A new insight was obtained by the discovery of a new glutathione peroxidase. This enzyme is smaller than the classical glutathione peroxidase, contains Se at the active site and reduces the phospholipid hydroperoxides to hydroxy derivatives (phospholipid hydroperoxide glutathione peroxidase) [24]. The peroxidase activity on membrane-bound hydroperoxides is apparently related to the observed dramatic inhibition of microsomal lipid peroxidation. The kinetic analysis showed that the reaction mechanism is the same for both peroxidases, but while glutathione peroxidase is more active on small peroxidic, substrates the new enzyme is more specific for lipidic 'interfacial' substrates [25].

The first enzymatic system against free radical damage (SOD and glutathione peroxidase) is devoted to the removal of non-lipid molecules possibly leading to primary initiations, the second system prevents

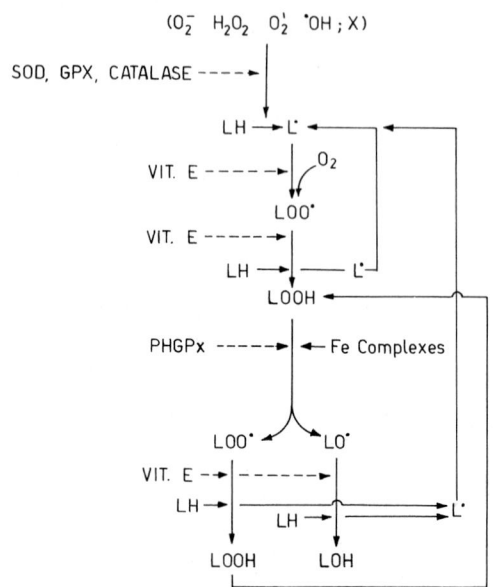

Fig. 1. Scheme of the multilevel system against lipid peroxidation. SOD = Superoxide dismutase; GPX = glutathione peroxidase; PHGPx = phospholipid hydroperoxide glutathione peroxidase; L = polyunsaturated phospholipid fatty acid.

secondary initiations. However, neither of these systems can control the chain propagation if lipid radicals are produced. A control at this level can be accomplished only by a free radical scavenger or by radical-radical chain-breaking reactions. The most common free radical scavenger in mammalian cells is the vitamin E [26]. The long debate on the antioxidant role of vitamin E over the past few years arose essentially from the low antioxidant activity in vitro. However, it is worth noting that in vivo the enzymes described above dramatically reduce the vitamin E requirement minimizing the free radical reactions. Moreover, recent evidences on the Se-dependent peroxidases give a rationale to the reciprocal sparing effect between vitamin E and Se [27]. Furthermore, the antioxidant capacity of vitamin E ist increased in the presence of ascorbate [28]. Following interaction with free radicals vitamin E is first oxidized to a radical and subsequently to tocopherylquinone. However, in the presence of ascorbate the first tocopheryl radical is reduced back to tocopherol and ascorbate is oxidized to dehydroascorbate that in turn disproportionates to ascorbate and dehydroascorbate [28]. As a consequence vitamin E content in the membranes is held constant and only ascorbate is consumed.

The scheme of this multilevel system against free radical damage is shown in figure 1.

References

1 Bulkey, G. B.: The role of oxygen free radicals in human disease processes. Surgery, St Louis *94:* 407–411 (1983).
2 Cutler, R. G.: Antioxidants, aging, and longevity; in Pryor, Free radicals in biology, vol.6, pp.371–428 (Academic Press, New York 1984).
3 McCord, J.; Fridovich, I.: Superoxide dismutase. J. biol. Chem. *244:* 6049–6055 (1969).
4 Bors, W.; Saran, M.; Tait, D. (eds): Oxygen radicals in chemistry and biology (de Gruyter, Berlin 1984).
5 Mead, J. F.: Free radical mechanisms of lipid damage and consequences for cellular membranes; in Pryor, Free radicals in biology, vol.1, pp.51–68 (Academic Press, New York 1976).
6 Tappel, A. L.: Measurement of lipid peroxidation from in vivo lipid peroxidation; in Pryor, Free radicals in biology, vol.4, pp.2–48 (Academic Press, New York 1980).
7 McCay, P. B.; King, M. M.; Poyer, J. L.: An update on antioxidant theory: spin trapping of trichloromethyl radicals in vivo. Ann. *1982:* 23–31.
8 Cadenas, E.; Bouveris, A.; Chance, B.: Low level chemiluminescence of biological systems; in Pryor, Free radicals in biology, vol.6, pp.212–243 (Academic Press, New York 1984).
9 Hayaishi, O. (ed.): Molecular mechanisms of oxygen activation (Academic Press, New York 1974).
10 Aust, S. D.; Morehouse, L. A.; Thomas, C. E.: Role of metals in oxygen radical reaction. J. Free Rad. Biol. Med. *1:* 3–25 (1985).
11 Fridovich, I.: Oxygen radicals, hydrogen peroxide, and oxygen toxicity; in Pryor, Free radicals in biology, vol.1, pp.239–277 (Academic Press, New York 1976).
12 Flohe, L.: Glutathione peroxidase brought into focus; in Pryor, Free radicals in biology, vol.5, pp.223–254 (Academic Press, New York 1982).
13 Pryor, W. A.: The role of free radical reactions in biological systems; in Pryor, Free radicals in biology, vol.1, pp.1–49 (Academic Press, New York 1976).
14 Green, M. J.; Hill, H. A. O.: Chemistry of dioxygen; in Packer, Methods in enzymology, vol.105, pp.3–22 (Academic Press, New York 1984).
15 Gabig, T. G.; Babior, B. M.: The superoxide forming oxidase responsible for the respiratory burst in human neutrophils. J. biol. Chem. *254:* 9070–9074 (1979).
16 McCord, J. M.: Oxygen-derived free radicals in post-ischemic tissue injury. New Engl. J. Med. *312:* 159–163 (1985).
17 Halliwell, B.: Superoxide-dependent formation of hydroxyl radicals in the presence of iron chelates. FEBS Lett. *92:* 321–326 (1978).
18 Chance, B.; Sies, H.; Boveris, A.: Hydroperoxide metabolism in mammalian organs. Physiol. Rev. *59:* 527–605 (1979).

19 Prohaska, J. R.: The glutathione peroxidase activity of glutathione S-transferases. Biochim. biophys. Acta *611:* 87–98 (1980).

20 Young, J. D.; Nimmo, I. A.; Hall, J. G.: The relationship between GSH, GSSG and non GSH-thiol in GSH deficent erythrocytes from Finnish Landrace and Tasmanian Merino sheep. Biochim. biophys. Acta *404:* 124 131 (1975).

21 Barsacchi, R.; Pelosi, G.; Camici, P.; Bonaldo, L.; Maiorino, M.; Ursini, F.: Glutathione depletion increases chemiluminescence emission and lipid peroxidation in the heart. Biochim. biophys. Acta *804:* 356–360 (1984).

22 Aust, S. D.; Svingen, B. A.: The role of iron in enzymatic lipid peroxidation; in Pryor, Free radicals in biology, vol. 5, pp. 1–28 (Academic Press, New York 1982).

23 Grossman, A.; Wendel, A.: Non-reactivity of the selenoenzyme glutathione peroxidase with enzymatically peroxidized phospholipids. Eur. J. Biochem. *135:* 549–552 (1983).

24 Ursini, F.; Maiorino, M.; Valente, M.; Ferri, L.; Gregolin, C.: Purification from pig liver of a protein which protects liposomes and biomembranes from peroxidative degradation and exhibits glutathione peroxidase activity on phosphatidyl choline hydroperoxides. Biochim. biophys. Acta *710:* 197–211 (1982).

25 Ursini, F.; Maiorino, M.; Gregolin, C.: The selenoenzyme phospholipid hydroperoxide glutathione peroxidase. Biochim. biophys. Acta *839:* 62–70 (1985).

26 Witting, L. A.: Vitamin E and lipid antioxidants in free radicals initiated reactions; in Pryor, Free radicals in biology, vol. 4, pp. 295–319 (Academic Press, New York 1980).

27 Diplock, A. T.; Lucy, Y. A.: The biochemical modes of action of vitamin E and selenium: a hypothesis. FEBS Lett. *29:* 205–210 (1973).

28 Scarpa, M.; Rigo, A.; Maiorino, M.; Ursini, F.; Gregolin, C.: Formation of tocopherol radical and recycling of ascorbate during peroxidation of phosphatidylcholine liposomes. Biochim. biophys. Acta *801:* 215–219 (1984).

F. Ursini, MD, Institute of Biological Chemistry, University of Padova, Via Marzolo 3, I-35100 Padova (Italy)

Novelli, Ursini (eds.), Oxygen Free Radicals in Shock. Int. Workshop, Florence 1985, pp. 15–28 (Karger, Basel 1986)

Mechanisms of Oxygen Free Radicals Production in Granulocytes[1]

Filippo Rossi, Vittorina Della Bianca, Miroslawa Grzeskowiak, Lucia Zeni

Istituto di Patologia Generale, Università degli Studi di Verona, Italy

In the last decade many data have been obtained in many laboratories showing that intermediates of O_2 reduction (O_2^-, OH·, H_2O_2 and other derivatives) are responsible for tissue injury in various pathological conditions [1–6]. Among the cellular systems generating O_2^- in different cell compartments (xanthine oxidase, aldehyde oxidase, flavoprotein dehydrogenase, mitochondrial and endoplasmic reticulum electron transport systems, peroxisomal enzymes, autoxidation of thiols, hemeprotein, drugs, etc.) the most relevant is the NADPH oxidase, the respiratory system of professional phagocytes (granulocytes, monocytes and macrophages). The activation of this system, which is located in the plasma membrane and is dormant in the resting state, results in an increase in O_2 consumption of phagocytes which is usually called 'respiratory burst'. Thus, the term respiratory burst refers to the increase in non-mitochondrial O_2 consumption, with concomitant production of O_2^-, H_2O_2, OH· and other toxic derivatives, and in the oxidation of glucose through the HMP shunt. The first observation was made by *Baldridge and Gerard* in 1933 [7] during phagocytosis by blood granulocytes. The phenomenon was extensively investigated from years 60 and in a short time it was clear that the respiratory burst is relevant for the defence against invading organisms, for the development of the inflammatory process and also for tissue injury [8–12].

We examine here the features, the mechanisms and, very briefly, the main effects of the respiratory burst.

[1] This work was supported by grants from Ministero P.I. (Fondo 40% Gruppo Difese Biologiche – coordinatore Prof. *Filippo Rossi*) and CNR (Contratto N. 84.00774.44 and Contributo N. 84.00789.04).

Table 1. The most common agents stimulating the oxidative metabolism of phagocyte

Phagocytosable particles	Phorbol esters
Adhesion to surfaces	N-Formyl peptides
Surfactants	Ca^{++} Ionophores
Anti-leukocyte antibody	Cytochalasin D, E
Phospholipase C	Platelet-activating factor
Immunocomplexes (Fc)	Leukotrienes
Complement fragments	Concanavalin A
Fatty acids	Diacylglycerol
Endotoxin	

(1) The respiratory burst was initially found associated with the phagocytosis but during the last two decades many data have been presented showing that a variety of factors are able to induce the activation of respiration in phagocytes (table I). It is worthy to mention that many of these factors are of biological relevance because they are frequently formed in the blood stream and in the inflammatory process. Thus, the problem arises whether the activation of the respiration in phagocytes is a frequent event in vivo.

(2) The respiratory burst is due to the activation of a plasma membrane bound oxidase which oxidises NADPH with formation of O_2^-. This free radical is released outside the cell or within the phagocytic vacuole where it rapidly dismutates to H_2O_2 (fig. 1) or can give origin to $OH^{\cdot}$ radical by interacting with H_2O_2, or to other oxidant species (hypochlorous acid, chloramines, aldehydes, etc.) in presence of myeloperoxidase secreted from the azurophylic granules.

(3) The respiratory burst can be measured as O_2 consumption; O_2^- and H_2O_2 production; $^{14}CO_2$ from $1\text{-}^{14}C$-glucose; nitroblutetrazolium reduction; chemiluminescence [for details see ref. 8].

(4) The respiratory burst is a very impressive phenomenon with regard the intensity and the rapidity of its occurrence. Experiments in vitro have shown that 1×10^7 human blood granulocytes can produce 80–100 nmol O^-_2/min in presence of maximal amount of phagocytosable bacteria or soluble stimulant, such as phorbol-myristate acetate (PMA). Depending on the type of the stimulant the burst starts after a lag time of 15 and 60 s and lasts few or many minutes. The lag time, the maximal rate, the duration vary depending on the stimulant, cell sources, priming state, dose of the stimulant and, in macrophages, on the functional state.

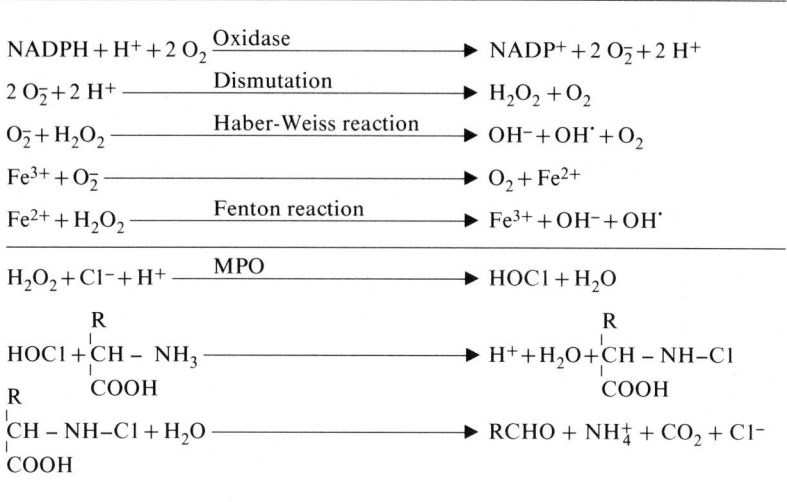

$$NADPH + H^+ + 2\,O_2 \xrightarrow{\text{Oxidase}} NADP^+ + 2\,O_2^- + 2\,H^+$$

$$2\,O_2^- + 2\,H^+ \xrightarrow{\text{Dismutation}} H_2O_2 + O_2$$

$$O_2^- + H_2O_2 \xrightarrow{\text{Haber-Weiss reaction}} OH^- + OH^{\cdot} + O_2$$

$$Fe^{3+} + O_2^- \longrightarrow O_2 + Fe^{2+}$$

$$Fe^{2+} + H_2O_2 \xrightarrow{\text{Fenton reaction}} Fe^{3+} + OH^- + OH^{\cdot}$$

$$H_2O_2 + Cl^- + H^+ \xrightarrow{\text{MPO}} HOCl + H_2O$$

$$HOCl + \underset{\underset{COOH}{|}}{\overset{\overset{R}{|}}{CH}} - NH_3 \longrightarrow H^+ + H_2O + \underset{\underset{COOH}{|}}{\overset{\overset{R}{|}}{CH}} - NH - Cl$$

$$\underset{\underset{COOH}{|}}{\overset{\overset{R}{|}}{CH}} - NH - Cl + H_2O \longrightarrow RCHO + NH_4^+ + CO_2 + Cl^-$$

Fig. 1. The reactive species produced during the respiratory burst.

(5) The respiratory burst is frequently associated with other functional responses such as aggregation, secretion, chemotaxis. One response is induced by many stimuli and one stimulus can produce many responses. This pluriresponsiveness is a relevant phenomenon for understanding the mechanisms of activation and their regulation.

Mechanism of the Respiratory Burst – the Enzymatic Basis

The enzyme (or the enzymatic system) responsible for the respiratory burst is the NADPH oxidase firstly described in 1964 in our laboratory [13, 14]. The oxidase is located in the plasma membrane with the active center directed towards the cytosol. In the course of phagocytosis only the enzyme of the portion of the membrane, which is invaginated and forms the phagosome, is activated [15], while when the stimulus is soluble the activation involves the oxidase distributed in all the plasma membrane.

The active oxidase catalyzes the oxidation of NADPH according the reaction $NADPH + H^+ + 2\,O_2 \rightarrow NADP^+ + 2\,O_2^- + 2\,H^+$. The k_m for NADPH is 0.02–0.05 mM, for NADH is 0.5–1.0 mM, for O_2 is about

0.03 mM. The optimum pH is 7.0–7.5. The activity is insensitive to KCN, rothenone, antimycin A, myxothiazol, and inhibited by parachlomercuribenzoate, cibacron blue, bathophenanthroline sulfonate, FAD analogues [16, 17].

The nature of the oxidase is under investigation in many laboratories. The main candidate component of the NADPH oxidase system are, at present, a FAD dehydrogenase directly involved in the subtraction of electrons from NADPH; a special type of cytochrome with a very negative (–245) mid-point redox potential [18, 19] and quinones [20]. We have solubilized and partially purified the oxidase from activated neutrophils [17, 21–24] by using a procedure that allows the separation of NADPH oxidase from NADH and NADPH dye reductase. The NADPH oxidase co-purified with cytochrome b_{-245}, was devoid of quinone and the FAD content varied in the various steps of the purification. In the crude solubilized extract the ratio cytochrome b/FAD was about 1:1 while in the most purified preparation the ratio was about 20:1 and even higher. During recent studies on the purification of the NADPH oxidase from activated neutrophils and macrophages we have identified a protein of 31,500 daltons which progressively increases during the various steps of extraction and purification of the enzyme [21, 23]. Recently we have reached evidences that this protein is the cytochrome b_{-245}. Furthermore, by using ^{32}P-labeled neutrophils both in resting and activated state, we have shown that this 31,500-dalton protein becomes phosphorylated when the cells are activated with PMA [24]. This finding could be relevant for understanding the mechanism of the activation of the NADPH oxidase and of the respiratory burst.

Mechanisms of Activation of the Oxidase

The overall phenomenon includes three phases: recognition, transduction, final process of activation.

The *first phase,* that is the interaction between the stimulus (particulate or soluble) and the cell surface, may involve receptors as in the case of Fc fragments of opsonized particles or immunocomplexes, C5a, formylated peptides, leukotriene B_4, lectins, etc., or binding molecules or substrates as in the case of phospholipase, detergents, fatty acids, phorbol esters. At the level of this phase many controls of the respiratory response are active. In general the activation of the respiratory burst is regulated by the occupancy, the rate of occupancy of the receptors, the affinity of the receptors, the number of receptors and a number of other events. We mention here some of these regulatory events in order to underline that the occurrence of the burst of production of O_2 radical by the phagocytes is a rather complex phenomenon.

(1) The affinity state of formyl-peptide receptors is regulated by guanine nucleotides (GTP, GDP), which transform the receptors originally present on leukocyte membrane in high-affinity state to low-affinity state [25].

(2) The respiratory response begins when only a fraction (10–20%) of the receptors is occupied [26, 27].

(3) The magnitude of the burst depends on the dose of the stimulant and also on the rate of presentation of the stimulus and therefore on the rate of the binding. When the presentation of the same dose of the stimulant is made over a period of minutes, instead than all at once, the respiratory response is greatly depressed. The slow rate of binding between stimulus and receptors generate an uncoupling between the binding and the apparatus for the response [27, 28].

(4) It has been found in our and in other laboratories that a continuous (new) binding of the stimulant is required for maintaining the respiration of phagocytes in activated state [29–32].

(5) A previous interaction with a stimulus generates a unresponsiveness (desensitization) to the same stimulus after the cessation of the first response. This fact is called 'down-regulation' and it is due to a decreased number of receptors available [33]. A previous treatment also with non-active doses of a stimulant can cause a hyperresponsiveness to another stimulant. This phenomenon is called 'priming' [34].

The *second phase* includes the series of molecular and functional modifications of the plasma membrane triggered by the recognition events. The main modifications likely involved in the activation of the burst are: an increase in Ca^{2+} influx [35, 36] and release from cell membranes and intracellular stores [37], with rapid increase in intracellular free Ca^{2+} concentration [38, 39]; an increase of turnover of phosphoinositides [40–42]; a translocation of proteinkinase C from the cytosol to plasma membrane [43]; an activation of protein phosphorylation [44–47]. The activation of Ca^{2+} and phospholipid-dependent proteinkinase C would be a common step, which ever is the stimulant, in the transduction reactions for leukocyte responses including the respiratory burst [48–51]. The main problems to be understood are the mechanism of activation of proteinkinase C, and the linking together the biochemical modifications that are associated with the functional responses of the phagocytes. In analogy with many other cells the messangers involved are Ca^{2+}, inositol triphosphate, diacylglycerol and phosphatidic acid. Results obtained very recently in many laboratories [57, 58] support the hypothesis that a guanine nucleotide binding protein mediates the coupling of chemotactic factors receptor occupancy to the other biochemical events and mediator production. Through a guanine nucleotide-binding protein, the agonist-receptor complex would induce the activation of a phospholipase C resulting in the hydrolysis of phosphatidylinositol 4,5-bisphosphate and formation of inositol triphosphate, which is responsible for the mobilization of Ca^{2+} from intracellular stores and opening of plasma membrane Ca^{2+} channels. Thus, the sequence of events would be as follows: ligand-receptor interaction; guanine nucleotide binding protein modification; activation of a membrane bound phospholipase C with formation of inositol triphosphate and increase in intracellular free Ca^{2+} $[Ca^{2+}]_i$; activation of a phospholipase D (or C) with hydrolysis of phosphatidylinositol and formation of phosphatidate and diacylglycerol; activation of the protein kinase C by diacyglycerol or other factors; phosphorylation of proteins and activation of NADPH oxidase (fig. 2). When the stimulant is PMA only the last step of the sequence of events takes place, since PMA directly activates the protein kinase C. Many points remain to be clarified, as for example the messangers responsible for the activation of phospholipase D, which could be the increase of intracellular $[Ca^{2+}]$.

In agreement with this hypothesis are the finding that: (a) after the treatment with FMLP there is an activation of turnover of phosphoinositides [40–42]. The activation regards two routes: one is the breakdown of phosphatidylinositol 4,5-bisphosphate which occurs at a very low dose of the stimulant and is responsible for the increase in $[Ca^{2+}]$; the other is the activation of a phospholipase D, which requires higher concentration of the stimulant, is responsible for the formation of diacylglycerol and phosphatidate and, through the activation of protein kinase C (or other mechanisms), of the functional response of the cell [59, 60]. (b) The activation is associated with a translocation of cytosol protein kinase C in plasma membrane [43]. (c) Diacylglycerol, an activator of proteinkinase C in vitro, stimulates the respiratory burst [49]. (d) The receptor of PMA, an activator of the respiratory burst, is the proteinkinase C [52]. (e) Unsaturated fatty acids activate the protein kinase in leukocyte homogenate [53] and the respiratory burst in intact leukocytes [30]. (f) The respiratory

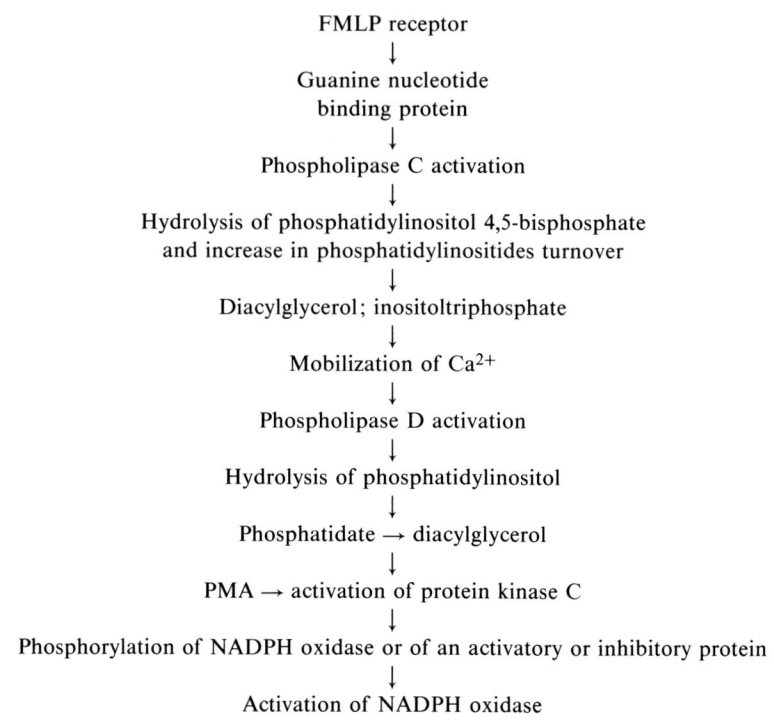

FMLP receptor
↓
Guanine nucleotide
binding protein
↓
Phospholipase C activation
↓
Hydrolysis of phosphatidylinositol 4,5-bisphosphate
and increase in phosphatidylinositides turnover
↓
Diacylglycerol; inositoltriphosphate
↓
Mobilization of Ca^{2+}
↓
Phospholipase D activation
↓
Hydrolysis of phosphatidylinositol
↓
Phosphatidate → diacylglycerol
↓
PMA → activation of protein kinase C
↓
Phosphorylation of NADPH oxidase or of an activatory or inhibitory protein
↓
Activation of NADPH oxidase

Fig. 2. Activation of NADPH oxidase. Possible sequence of events.

burst induced by exogenous phospholipase C is associated with a production of diacylglycerol [54].

The *third phase* is the activation of the terminal respiratory system.
On the basis of the hypothesis of the involvement of the proteinkinase C,
the transition of the respiratory system from inactive to active state
would be preceded by the translocation of the proteinkinase C from the
cytosol to the membrane and would be triggered by a process of phosphorylation of some component of the oxidase. It is worthy pointing out
that we have recently demonstrated that the 31,500-dalton protein corresponding to the cytochrome b_{-245} is phosphorylated when extracted from
granulocytes activated by PMA [24]. The relationship between the phosphorylation process and the activated state remain to be investigated. It
is likely that the phosphorylation causes a conformational change of one

component of the oxidase resulting in a modification of the kinetic properties (exposition of active group, modification of K_m for the substrate) or in a facilitation in the assembling of the components of the oxidase [55]. A clarification of this matter will be possible when the precise nature of the oxidase (multicomponent or one component system) is understood.

Effects of the Respiratory Burst

The effects are linked to the production fate and reactivity of oxygen free radicals, of H_2O_2 and of the other species indirectly formed (hypochlorous acid, chloramines, aldehydes). These compounds act mainly as oxidant and interact with many molecular species inside and outside the cell (table II). By virtue of these reactions the products of the burst play a very important *defensive* role (killing of invading organisms and tumor cells), *modulate the development of the inflammatory process, and are responsible for many tissues damages and pathological conditions* (degeneration at acute and chronic inflammatory sites, lung injury in adult respiratory distress syndrome; part of tissue damage during shocks and postischemic reperfusion, etc.). Apart from a detailed analysis of the defensive, toxic and modulatory effect, which have been made elsewhere [8, 11, 56], we believe that some points are worthy of discussion in the contest of a meeting on oxygen free radicals and shock.

Table II. Harmful effect of oxygen free radicals

Target	Consequence
Thiol-containing amino acids	protein denaturation, cross-linking, polimerization, degradation, enzyme inhibition
Nucleic acids	cell-cycle change, mutation, strand scission, base modification
Carbohydrate	cell surface receptor changes
Lipids	peroxidation, organelle and membrane changes
Cofactors	metabolic changes
Hyaluronic acid	depolimerization
Biological factors	inactivation (i.e. α_1-antitrypsin, chemotactic factors, mediators, neurotransmitters)

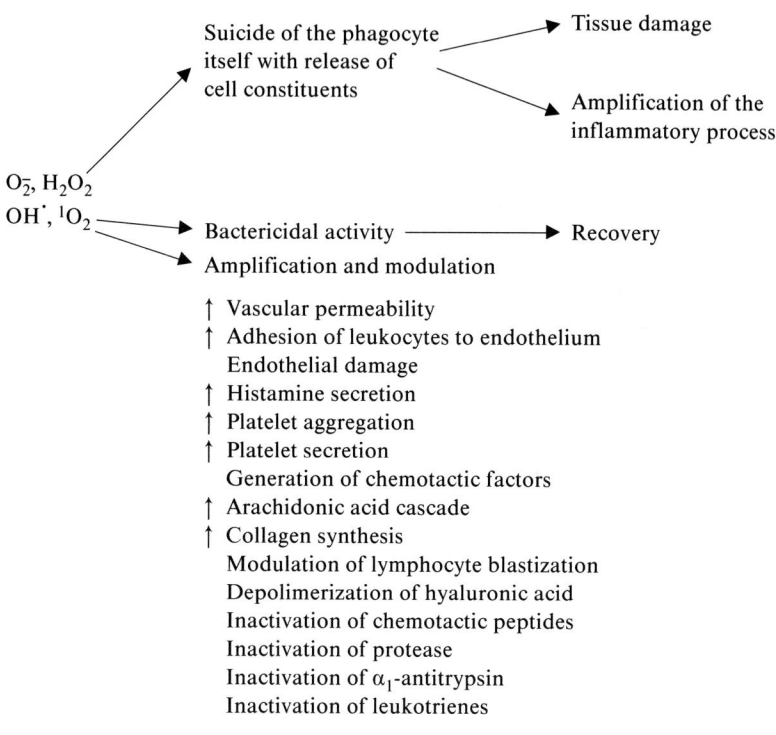

O_2^-, H_2O_2
$OH^{\cdot}$, $^1O_2$

Suicide of the phagocyte itself with release of cell constituents → Tissue damage

→ Amplification of the inflammatory process

Bactericidal activity ———→ Recovery

Amplification and modulation

↑ Vascular permeability
↑ Adhesion of leukocytes to endothelium
   Endothelial damage
↑ Histamine secretion
↑ Platelet aggregation
↑ Platelet secretion
   Generation of chemotactic factors
↑ Arachidonic acid cascade
↑ Collagen synthesis
   Modulation of lymphocyte blastization
   Depolimerization of hyaluronic acid
   Inactivation of chemotactic peptides
   Inactivation of protease
   Inactivation of α_1-antitrypsin
   Inactivation of leukotrienes

Fig. 3. Respiratory burst and inflammatory process.

The criteria for the involvement of oxygen radical in different processes (defensives, toxic, etc.) are the following: (a) the effects or the pathological processes should be inhibited by superoxide dismutase, catalase and scavengers; (b) the production of free radicals or H_2O_2 should be directly demonstrated; (c) the production of toxic compounds derived from reactions triggered by oxygen radicals or by H_2O_2 should be demonstrated; (d) many effects or the pathological processes should be reproduced by oxygen radicals or H_2O_2-producing systems.

In our opinion many evidences presented by many laboratories are indirect and not sufficiently proved. In fact the only demonstration of the conditions for the activation of the respiratory burst (for example the formation of C5a in the blood, the aggregation of neutrophils or their presence in the damaged tissues, etc.) does not mean that the leukocytes are actually in activated state, that is that they are producing toxic oxygen species. This should be directly demonstrated as indicated in the criteria b and c. As briefly discussed before the occurrence, magnitude and duration of the respiratory burst are regulated in a very complex manner by many factors. It is clear that many of these regulatory factors act as modulator or depressor of a process which is functional to defensive mechanisms (bactericidal, tumoricidal) and which is potentially dangerous for the tissue and also for the leukocyte itself. We are of the opinion that the induction of the respiratory burst, that is of the production of the free radical of oxygen in leukocytes, is not so easy in vivo as in vitro.

Table II reports the harmful effects of oxygen free radicals and figure 3 the relationship between the respiratory burst and the inflammatory process. The defensive activity (bactericidal, virocidal, tumoricidal) is exerted inside the phagosome or outside the cells. Two mechanisms have been demonstrated: a direct oxidizing action of free radicals and H_2O_2 and, more relevant, the generation of toxis species mediated by myeloperoxidase, H_2O_2 and halide [7].

References

1 Slater, T. F.: Free-radical mechanism in tissue injury. Biochem. J. *222:* 1–15 (1984).
2 Halliwell, B.; Gutteridge, J. M. C.: Oxygen toxicity, oxygen radicals, transition metals and disease. Biochem. J. *219:* 1–14 (1984).
3 McCord, J. M.: Oxygen-derived free radicals in postischemic tissue injury. New Engl. J. Med. *312:* 159–163 (1985).
4 Fridovich, I.: Superoxide radical: an endogenous toxicant. Annu. Rev. Pharmacol. Toxicol. *23:* 239–257 (1983).
5 Cohen, G.; Greenwald, R.: Oxy radicals and their scavanger systems, vol. I, II (Elsevier Biomedical, New York 1982).
6 Trush, M. A.; Mimnaugh, E. G.; Gram, T. E.: Activation of pharmacologic agents to radical intermediates. Implications for the role of free radicals in drug action and toxicity. Biochem. Pharmacol. *31:* 3335–3346 (1982).
7 Baldridge, C. W.; Gerard, R. W.: The extrarespiration of phagocytosis. Am. J. Physiol. *103:* 235–236 (1933).
8 Klebanoff, S. J.; Clark, R. A.: The neutrophil: function and clinical disorders (North-Holland, Amsterdam 1978).
9 Rossi, F.; Patriarca, P.; Romeo, D.: Metabolic changes accompanying phagocytosis; in Sbarra Strauss, The reticuloendothelial system, vol. 2, pp. 153–188 (Plenum Publishing, New York 1980).

10   Badway, J. A.; Karnovsky, M. L.: Active oxygen species and the functions of phagocytic leukocytes. A. Rev. Biochem. *49:* 695–726 (1980).

11   Weiss, S. J.; Lo Buglio, A. F.: Biology of disease. Phagocyte-generated oxygen metabolites and cellular injury. Lab. Invest. *47:* 5–18 (1982).

12   Babior, B. M.: Oxygen-dependent microbial killing by phagocytes. New Engl. J. Med. *298:* 659–668 (1978).

13   Rossi, F.; Zatti, M.: Changes in the metabolic pattern of polymorphonuclear leukocytes during phagocytosis. Br. J. exp. Path. *45:* 548–559 (1964).

14   Rossi, F.; Zatti, M.: Biochemical aspects of phagocytosis in polymorphonuclear leucocytes. NADH and NADPH oxidation by the granules of resting and phagocytosing cells. Experentia *20:* 21–24 (1964).

15   Bellavite, P.; Serra, M. C.; Davoli, A.; Rossi, F.: Selective enrichment of NADPH oxidase activity in phagosomes from guinea pig polymorphonuclear leucocytes. Inflammation *6:* 21–29 (1982).

16   Light, D. R.; Walsh, C.; O'Callaghan, A. M.; Goetzl, E. J.; Tauber, A. I.: Characteristics of the cofactor requirements for the superoxide-generating NADPH oxidase of human polymorphonuclear leukocytes. Biochemistry, N. Y. *20:* 1468–1476 (1981).

17   Bellavite, P.; Serra, M. C.; Davoli, A.; Bannister, J. V.; Rossi, F.: The NADPH oxidase of guinea pig polymorphonuclear leucocytes. Properties of the deoxycholate extracted enzyme. Mol. cell. Biochem. *52:* 17–25 (1983).

18   Shinagawa, Y.; Tanaka, C.; Teraoka, A.; Shinagawa, Y.: A new cytochrome in neutrophil granules of rabbit leukocytes. J. Biochem. *59:* 622–624 (1966).

19   Segal, A. W.: Chronic granulomatous disease: a model for studying the role of cytochrome b_{-245} in health and disease; in Gallin, Fauci, Granulomatous disease, pp. 121–143, (Raven Press, New York 1983).

20   Crawford, D. R.; Schneider, D. L: Identification of ubiquinone-50 in human neutrophils and its role in microbicidal events. J. biol. Chem. *257:* 6662–6668 (1982).

21   Bellavite, P.; Jones, O. T. G.; Cross, A. R.; Papini, E.; Rossi, F.: Comparison of partially purified NADPH oxidase from pig neutrophils. Biochem. J. *223:* 639–648 (1984).

22   Serra, M. C.; Bellavite, P.; Davoli, A.; Bannister, J. V. ; Rossi, F.: Isolation from neutrophil membranes of a complex containing active NADPH oxidase and cytochrome b_{-245}. Biochim. biophys. Acta *788:* 138–146 (1984).

23   Rossi, F.; Bellavite, P.; Serra, M. C.; Papini, E.: Characterization of phagocyte NADPH oxidase; in Van Furth, Mononuclear phagocytes. Characteristics, physiology and function, pp. 423–433 (Martinus Nijhoff, Dordrecht 1985).

24   Bellavite, P.; Papini, E.; Zeni, L.; Della Bianca, V.; Rossi, F.: Studies on the nature and activation of O_2^- forming NADPH oxidase of leucocytes. Identification of a phosphorylated component of the active enzyme. Free Radicals Res. Commun. *1:* 11–29 (1985).

25   Snyderman, R.: Regulatory mechanisms of a chemoattractant receptor on leucocytes. Fed. Proc. *43:* 2743–2748 (1984).

26   Painter, R. G.; Sklar, L. A.; Jesaitis, A. J.; Schmitt, M.; Cochrane, C. G.: Activation of neutrophils by N-formyl chemotactic peptides. Fed. Proc. *43:* 2737–2742 (1984).

27   De Togni, P.; Bellavite, P.; Della Bianca, V.; Grzeskowiak, M.; Rossi, F.; Inten-

sity and kinetics of respiratory burst of human neutrophils in relation to receptor occupancy and rate of occupation by formylmethionylleucylpenylalanine. Biochim. biophys. Acta *838:* 12–22 (1985).

28 Sklar, L. A.; Jesaitis, A. J.; Painter, R. G.; Cochrane, C. G.: The kinetics of neutrophil activation. The response to chemotactic peptides depends upon whether ligand-receptor interaction is rate-limiting. J. biol. Chem. *256:* 9909–9914 (1981).

29 Romeo, D.; Zabucchi, G.; Rossi, F.: Reversible metabolic stimulation of polymorphonuclear leukocytes and macrophages by concanavalin A. Nature, Lond. *243:* 111–112 (1973).

30 Badwey, J. A.; Curnutte, J. T.; Robinson, J. M.; Berde, C. B.; Karnovsky, M. J.; Karnovsky, M. L.: Effects of free fatty acids on release of superoxide and on change of shape by human neutrophils. Reversibility by albumin. J. biol. Chem. *259:* 7870–7877 (1984).

31 Rossi, F.; De Togni, P.; Bellavite, P.; Della Bianca, V.; Grzeskowiak, M.: Relationship between the binding of *N*-formylmethionylleucylphenylalanine and the respiratory response in human neutrophils. Biochim. biophys. Acta *758:* 168–175 (1983).

32 Sklar, L. A.; Oades, Z. G.; Jesaitis, A. J.; Painter, R. G.; Cochrane, C. G.; Fluoresceinated chemotactic peptide and high-affinity antifluorescein antibody as a probe of the temporal characteristics of neutrophil stimulation. Proc. natn. Acad. Sci. USA *78:* 7540–7544 (1981).

33 Donabedian, H.; Gallin, J. I.: Deactivation of human neutrophil chemotaxis by chemoattractants: effect on receptors for the chemotactic factor f-met-leu-phe. J. Immun. *127:* 839–844 (1981).

34 McPhail, L. C.; Clayton, C. C.; Snyderman, R.: The NADPH oxidase of human polymorphonuclear leukocytes. Evidence for regulation by multiple signals. J. biol. Chem. *259:* 5768–5775 (1984).

35 Naccache, P. H.; Showell, H. J.; Becker, E. L.; Sha'afi, R. I.: Changes in ionic movements across rabbit polymorphonuclear leukocyte membranes during lysosomal enzyme release. Possible ionic basis for lysosomal enzyme release. J. biol. Chem. *75:* 635–649 (1977).

36 Korchak, H. M.; Rutherford, E.; Weissmann, G.: Stimulus response coupling in the human neutrophil. I. Kinetic analysis of changes in calcium permeability. J. biol. Chem. *259:* 4070–4075 (1984).

37 Weissmann, G.; Hoffstein, S.; Korchak, H.; Smolen, J. E.: The earliest membrane responses to phagocytosis: membrane potential changes and Ca^{2+} loss in human granulocytes. Trans. Am. Ass. Physns *91:* 90–101 (1983).

38 Pozzan, T.; Lew, D. P.; Wollheim, C. B.; Tsien, R. Y.: Is cytosolic ionized calcium regulating neutrophil activation? Science *221:* 1413–1415 (1983).

39 White, J. R.; Naccache, P. H.; Molski, T. F. P.; Borgeat, P.; Sha'afi, R. I.: Direct demonstration of increased intracellular concentration of free calcium in rabbit and human neutrophils following stimulation by chemotactic factor. Biochem. biophys. Res. Commun. *113:* 44–50 (1983).

40 Serhan, C. N.; Broekman, M. J.; Korchak, H. M.; Smolen, J. E.; Marcus, A. J.; Weissmann, G.: Changes in phosphatidylinositol and phosphatidic acid in stimulated human neutrophils. Relationship to calcium mobilization, aggregation and superoxide radical generation. Biochim. biophys. Acta *762:* 420–428 (1983).

41 Daugherty, R. W.; Godfrey, P. P.; Hoyle, P. C.; Putney, J. M. Jr.; Freer, R. J.: Secretagogue-induced phosphoinositide metabolism in human leucocytes. Biochem. J. *222:* 307–314 (1984).

42 De Togni, P.; Della Bianca, V.; Grzeskowiak, M.; Di Virgilio, F.; Rossi, F.: Mechanism of desensitization of neutrophil response to N-formylmethionylleucylphenylalanine by slow rate of receptor occupancy. Studies on changes in Ca^{2+} concentration and phosphatidylinositol turnover. Biochim. biophys. Acta *838:* 23–31 (1985).

43 McPhail, L. C.; Wolfson, M.; Clayton, C.; Snyderman, R.: Protein kinase C and neutrophil (PMN) activation. Differential effects of chemoattractants and phorbol myristate acetate (PMA). Fed. Proc. *43:* 1661, abstr. 1430 (1984).

44 Andrews, P. C.; Babior, B. M.: Endogenous protein phosphorylation by resting and inactivated human neutrophils. Blood *61:* 333–340 (1983).

45 Irita, K.; Takeshige, K.; Minakami, S.: Protein, phosphorylation in intact pig leukocytes. Biochim. biophys. Acta *805:* 44–52 (1984).

46 Schneider, C.; Zanetti, M.; Romeo, D.: Surface-reactive stimuli selectively increase protein phosphorylation in human neutrophils. FEBS Lett. *127:* 4–8 (1981).

47 Okamura, N.; Ohashi, S.; Nagahisa, N.; Ishibashi, S.: Changes in protein phosphorylation in guinea pig polymorphonuclear leukocytes by treatment with membrane-pertubing agents which stimulate superoxide anion production. Archs Biochem. Biophys. *228:* 270–277 (1984).

48 Sha'afi, R. I.; White, J. R.; Molsky, T. F. P.; Shefcyk, J.; Volpi, M.; Naccache, P. H.; Feinstein, M. B.: Phorbol 12-myristate 13-acetate activates rabbit neutrophils without an apparent rise in the level of intracellular free calcium. Biochim. biophys. Res. Commun. *114:* 638–645 (1983).

49 Fujta, I.; Irita, K.; Takeshige, K.; Minakami, S.: Diacylglycerol, 1-oleoyl-2acetylglycerol, stimulates superoxide-generation from neutrophils. Biochem. biophys. Res. Commun. *120:* 318–324 (1984).

50 Di Virgilio, F.; Lew, D. P.; Pozzan, T.: Protein kinase C activation of physiological processes in human neutrophils at vanishingly small cytosolic Ca^{2+} levels. Nature, Lond. *310:* 691–694 (1984).

51 Helfman, D. M.; Appelbaum, B. D.; Vogler, W. R.; Kuo, J. F.: Phospholipid-sensitive Ca^{2+}-dependent protein kinase and its substrates in human neutrophils. Biochem. biophys. Res. Commun. *111:* 847–853 (1983).

52 Castagna, M.; Takai, Y.; Kaibuchi, K.; Sano, K.; Kikkawa, U.; Nishizuka, Y.: Direct activation of calcium-activated, phospholipid-dependent protein kinase by tumor-promoting phorbol esters. J. biol. Chem. *257:* 7847–7851 (1982).

53 McPhail, L. C.; Clayton, C.; Snyderman, R. A.: A potential second messenger role for unsatured fatty acids: activation of Ca^{2+}-dependent protein kinase. Science *224:* 622–624 (1984).

54 Grzeskowiak, M.; Della Bianca. V.; De Togni, P.; Papini, E.; Rossi, F.: Independence with respect to Ca^{2+} changes of the neutrophil respiratory and secretory response to exogenous phospholipase C and possible involvement of diacyglycerol and protein kinase C. Biochim. biophys. Acta *844:* 81–89 (1985).

55 Borregaard, N.; Heiple, J. M.; Simons, E. R.; Clark, R. A.: Subcellular localization of the b-cytochrome component of the human neutrophil microbicidal oxidase: translocation during activation. J. Cell Biol. *97:* 52–61 (1983).

56 Rossi, F.; Dri, P.; Bellavite, P.; Zabucchi, G.; Berton, G.: Oxidative metabolism of inflammatory cells; in Weissmann, Samuelsson, Paoletti, Advances in inflam. res., vol. I, pp. 139–156 (Raven Press, New York 1979).

57 Cockcroft, S.; Gomperts, B. D.: Role of guanine nucleotide binding protein in the activation of phosphoinositide phosphodiesterase. Nature, Lond. *314:* 534–536 (1985).

58 Shefcyk, J.; Yassin, R.; Volpi, M.; Molski, P. H.; Naccache, P. H.; Munor, J. J.; Becker, E. L.; Feinstein, M. B.; Sha'afi, R. I.: Pertussis but not cholera toxin inhibits the stimulated increase in action association with the cytoskeleton in rabbit neutrophils: role of the 'G protein' in stimulus-response coupling. Biochem. biophys. Res. Commun. *126:* 1174–1181 (1985).

59 Grzeskowiak, M.; Della Bianca, V.; De Togni, P.; Cassatella, M.; Rossi, F.: The reactions of phosphoinositide response involved in Ca^{2+} changes and activation of NADPH oxidase in human neutrophils treated with fMet-Leu-Phe. Biochim. biophys. Acta (submitted for publication).

60 Cockcroft, S.; Borrowman, B. B.; Gomperts, B. D.: Breakdown and synthesis of polysphoinositides in fMetLeuPhe-stimulated neutrophils. FEBS Lett. *181:* 259–263 (1985).

F. Rossi, MD, Istituto di Patologia Generale, Università degli Studi di Verona, Strada Le Grazie, I-37134 Verona (Italy)

Novelli, Ursini (eds.), Oxygen Free Radicals in Shock. Int. Workshop, Florence 1985, pp. 29–33 (Karger, Basel 1986)

Determination of Reduced and Oxidized Ubiquinones and Tocopherol in Human Muscular Tissue[1]

M. Antonelli[a], G. G. Corbucci[a], M. Bufi[a], R. A. De Blasi[a], A. Gasparetto[a], O. Ghirardi[b], A. Peschechera[b]

[a]Anaesthesia and Resuscitation Institute, University La Sapienza;
[b]Biological Research Laboratories, Sigma Tau Pomezia, Rome, Italy

Introduction

The ubiquinones are essential components of mitochondria. This family of lipid-soluble benzoquinones acts as electron and proton carriers in the respiratory chain, representing a mobile pool between dehydrogenases and cytochromes [1, 2]. Furthermore, quinones appear to have other functions, such as an antioxidant effect [3–5] or a role in membrane fluidity [6, 7].

The amount of quinones in plasma samples was already investigated using high-performance liquid chromatography (HPLC) combined with an electrochemical detector (ECD) and an ultraviolet spectrometric detector (UVD) [8, 9].

We determined a quantitative analysis of ubiquinones in human muscular tissue by HPLC combined with an ECD, in dual mode, with the electrodes assembled in series.

---

[1] This work conforms to ethical considerations for research on hospital patients and was approved by the Human Experimental Review Committee of the Anaesthesia and Resuscitation Department, University of Rome.

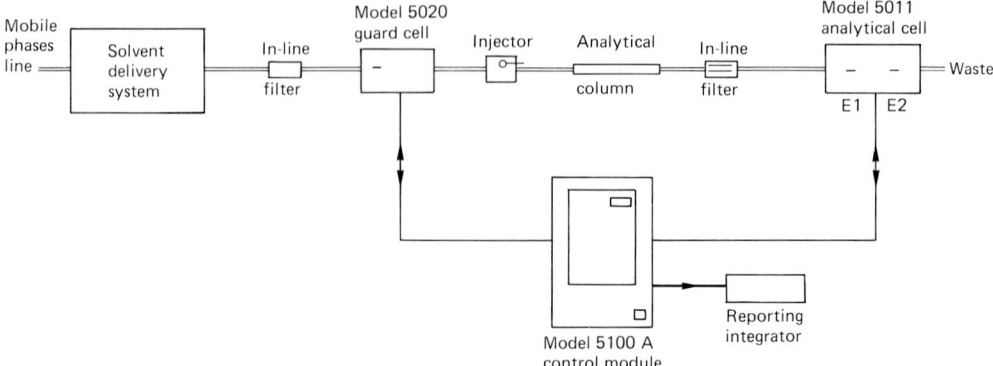

Fig. 1. The HPLC system.

Materials and Methods

Apparatus. The HPLC-ECD consisted of a Waters Pump 6000A with a Waters U6K loop injector, a Coulochem ESA 5100A control module, with Guard Cell No. 5020 (placed before the injector) and a high sensitivity analytical cell (ESA) with two porous graphite in-line working electrodes (fig. 1). A reverse-phase column Lichrosorb RP 18 (5 µm), 125×4.6 mm i.d. Merk Dermstadt was adopted. The mobile phase was prepared by dissolving 2 g of NaOH in 1,000 ml of a solution of ethanol-methanol-70% $HClO_4$ (500:500:5.3). The HPLC measurements were performed at room temperature. The flow-rate was 1 ml/min. In order to prevent the oxidation by dissolved oxygen during separation, the mobile phase and all extraction reagents were de-aerated by helium gas bubbling. HP3390A was used for graphic integration.

Chemicals. Ubiquinone-7 (Q_7), ubiquinone-9 (Q_9), ubiquinone-10 (Q_{10}), and α-tocopherol acetate (vitamin E) were synthesized by the Sigma Chemical Company. Ubiquinol 10 ($Q_{10}H_2$) was obtained by the reduction of Q_{10} with sodium borohydride [8, 9]. Other chemicals (reagent grade) were used without further purification.

Extraction. Needle biopsy samples from vastus lateralis muscles [10] of 16 volunteers and 10 shock patients were taken, frozen immediately in liquid nitrogen and stored at −40 °C for about 4 weeks before analysis. The muscular tissue, 60±20 mg average wet weight, was homogenized at 5 °C with 1 ml of 10 mM Tris-hcl (pH 7.4)/0.25 M sucrose/5 mM $MgCl_2$/0.1 mM ethylenediaminetetraacetic acid (EDTA)/200 mM sodium dodecyl sulphate (SDS) [11] in an Ultra-Terrax TP 18/10 (JK) homogenizer.

Into the text tube was added 2 ml of a mixture of ethanol-*n*-exane (2:5 by volume); the tube was rapidly shaken for 5 min to extract the ubiquinones. The solution was then centrifuged for 10 min at 2,000–2,200 *g* in order to separate the two layers. The extraction was repeated twice. The combined *n*-exane layer was dried under a stream of helium and resuspended in 0.3 ml of ethanol, stored at −80 °C for about 15 h and then subjected to HPLC. All the extraction steps were performed at 4 °C, in the absence of direct sunlight.

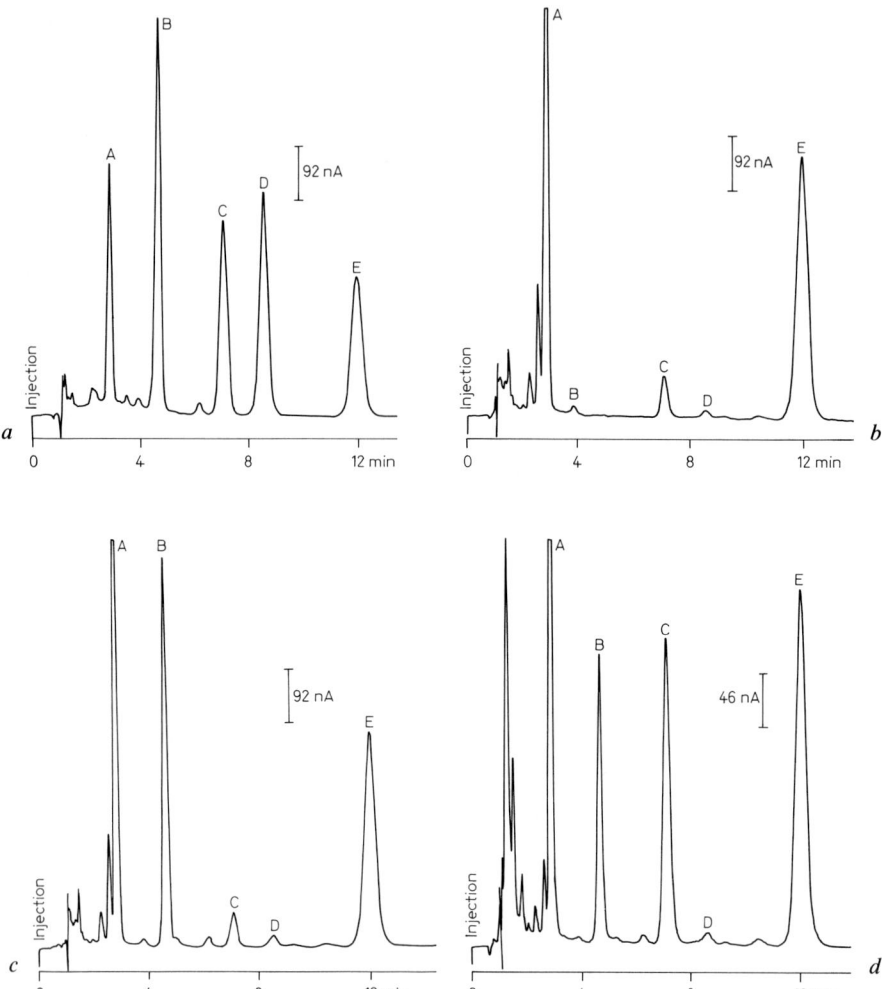

Fig. 2. Chromatograms of vitamin E, reduced and oxidized ubiquinones. Column, Lichrosorb RP 18 (5 μm), 125×4.6 mm i.d.; mobile phase ethanol-methanol 70% $HClO_4$ (500:500:5.3), containing 50 mM NaOH; flow rate, 1.0 ml/min; detection, ED, −0.6 V detector 1, +0.6 V detector 2. *a* Standards (amount injected: 60 ng of each Q, 46 μg of vitamin E). *b* Normal subject without internal standard. *c* Normal subject with internal standard. *d* Shock patient.

Peak A, vitamin E, t_R 2.80 min; peak B, Q_7, t_R 4.60 min; peak C, $Q_{10}H_2$, t_R 7.00 min; peak D, Q_9, t_R 8.60 min; peak E, Q_{10}, t_R 12.00 min.

Results

Identification of vitamin E, Q_9, Q_{10} and $Q_{10}H_2$ was based on comparisons of retention times of chromatographic peaks with the respective standards. Guard-cell potential was -0.8 V; applied reading-cell potentials were -0.6 for detector 1 and $+0.6$ for detector 2. With this system the reduced forms passed through detector 1 not altered, to be read through detector 2. The oxidized forms were reduced from detector 1 and reoxidized from detector 2. Q_7 was used as internal standard because the muscular tissue did not show any interference peak at Q_7 retention time. The retention times are shown in figure 2.

Q_9, Q_{10}; $Q_{10}H_2$ and vitamin E were clearly separated from each other and had retention times consistent with those of authentic compounds. Extractability was constant in the range of 95–98%. The calibration curve was linear from 5 to 200 ng injected of ubiquinones and from 50 to 500 µg injected of vitamin E.

Discussion

The method reported provided a rapid, sensitive and direct determination and separation of the oxidized and reduced uniquinones and vitamin E in human muscular biopsies.

The in-line working electrodes obtained a *simultaneous determination of* Q_{ox} *and* Q_{red} *simply using HPLC-ECD without the UVD*. The SDS permitted a better extraction of quinones [11]. The comparison of data from normal subjects with those from shock patients was very interesting (table I). In fact an increase of vitamin E, of $Q_{10}H_2$ and a decrease of Q_{10},

Table I. Levels of Q_{10}, $Q_{10}H_2$, Q_9 and vitamin E in human muscular tissue

Normal subjects (n = 16; $\bar{x} \pm$ SEM) $\mu g \cdot g^{T-1}$						Shock patients (n = 10; $\bar{x} \pm$ SEM) $\mu g \cdot g^{T-1}$					
vit. E $\mu g \cdot g^{T-1}$	Q_{10}	$Q_{10}H_2$	Q_{10TOT}	$Q_{10}/Q_{10}H_2$	Q_9	vit. E $\mu g \cdot g^{T-1}$	Q_{10}	$Q_{10}H_2$	Q_{10TOT}	$Q_{10}/Q_{10}H_2$	Q_9
15.09 ±1.35	34.65 ±3.20	6.48 ±0.79	41.38 ±3.16	7.15 ±1.75	0.996 ±0.194	31.74 ±6.04	15.29 ±1.70	11.19 ±3.29	26.48 ±2.07	4.99 ±2.07	0.454 ±0.063

Q_{10tot} and Q_9 during shock were observed. A relationship with cellular and mitochondrial anoxic damage is suggested.

References

1 Folkers, K.; Yamamura, Y.: Biomedical and clinical aspects of coenzyme Q, No. 1–4 (Elsevier, Amsterdam 1977–1984).

2 Lenaz, G.; Parenti-Castelli, G.: Multiple role of ubiquinone in mammalian cells. Drugs exp. clin. Res. *10:* 481–490 (1984).

3 Meilors, A.; Tappel, A. L.: The inhibition of mitochondrial peroxidation by ubiquinone and ubiquinol. J. biol. Chem. *241:* 4353 (1966).

4 Landi, L.; Cabrini, L.; Sechi, A. M.; Pasquali, P.: Antioxidative effect of ubiquinones on mitochondrial membranes. Biochem. J. *222:* 463–466 (1984).

5 Takeshige, K.; Takayanagy, K.; Minakami, S.: Reduced coenzyme Q_{10} as an antioxidant of liquid peroxidation in bovine heart mitochondria; in Yamamura, Folkers, Biomedical and clinical aspects of coenzyme Q, No. 2, pp. 15–26 (Elsevier, Amsterdam 1980).

6 Katsikas, H.; Quinn, P. J.: The polyisoprenoid chain length influences the interaction of ubiquinones with phospholipid bilayers. Biochim. biophys. Acta *689:* 363 (1982).

7 Lenaz, G.; Curatola, G.; Fiorini, R. M.; Parenti-Castelli, G.: Membrane fluidity and its role in the regulation of cellular processes; in Mirand, Hutchinson, Mihich, Biology of cancer, No. 2, pp. 25–34 (Liss, New York 1983).

8 Ikenoya, S.; Takada, M.; Yuzuriha, T.: Abe, K.; Katayama, K.: Studies on reduced and oxidized ubiquinones. 1. Simultaneous determination of reduced and oxidized ubiquinones in tissue and mitochondria by high performance liquid chromatography. Chem. pharm. Bull., Tokyo *29:* 158–164 (1981).

9 Takada, M.; Ikenoya, S.; Yuzuriha, T.; Katayama, K.: Simultaneous determination of reduced and oxidized ubiquinones; in Packer, Oxigen radicals in biological system. Methods in enzymology, No. 105, pp. 147–155 (Academic Press, Orlando 1984).

10 Edwards, R.; Young, A.; Wiles, M.: Needle biopsy of skeletal muscle in the diagnosis of myopathy and clinical study of muscle functions and repair. New Engl. J. Med. *302:* 261–271 (1980).

11 Hirota, K.; Kawase, M.: Effect of sodium dodecyl sulphate on the extraction of ubiquinone 10 in the determination of plasma sample. J. Chromatogr., biomed. Appl. *310:* 204–207 (1984).

M. Antonelli, MD, Institute of Anaesthesia and Resuscitation, University La Sapienza, Policlinico Umberto I°, Viale del Policlinico, I-00161 Rome (Italy)

Novelli, Ursini (eds.), Oxygen Free Radicals in Shock. Int. Workshop,
Florence 1985, pp. 34–41 (Karger, Basel 1986)

Lipid Peroxidation Products in Experimental Thermal Injury[1]

Gerd O. Till, John R. Hatherill, Peter A. Ward

Department of Pathology, University of Michigan Medical School,
Ann Arbor, Mich., USA

Experimental and clinical studies have shown that thermal injury to
skin can result in activation of the complement system producing a pattern that appears to signal activation of both the alternative and classical
pathways [1, 2]. Although the mechanism(s) of complement activation
after skin burns is not yet known, the possibility that this is due to heat-
denatured plasma proteins, cells or tissues has been discussed [3]. Employing at rat model of skin thermal injury (70 °C, 30 s), we have observed systemic complement activation which is accompanied by the appearance in plasma of C5-derived chemotactic activity for neutrophils,
concomitant transient neutropenia and sequestration of blood neutrophils in pulmonary capillaries [2]. These events are followed, at 2–3 h
after thermal injury, by significant increases in lung vascular permeability (as determined by transudation into the lung interstitium) of intravascular ^{125}I-bovine serum albumin and morphological changes showing
neutrophils and damaged endothelial cells in pulmonary capillaries, and
the presence of interstitial and intraalveolar edema and hemorrhage [2].
Because acute lung injury secondary to skin burns can almost completely
be prevented in rats previously depleted of either complement or circulating neutrophils and because the injury is also dramatically attenuated in
animals pretreated with antioxidant enzymes (catalase, superoxide dismutase), iron chelators (apolactoferrin, deferoxamine mesylate, 2,3-dihy-
droxybenzoic acid) or scavengers of hydroxyl radical (dimethyl sulfoxide, dimethyl thiourea, sodium thiourea, sodium benzoate) [2, 4] it is believed that oxygen radicals released from complement-activated neutro-

[1] Supported in part by NIH Grants GM28499 and GM29507.

phils are the most likely mediators of acute lung microvascular injury. This assumption is strongly supported by similar findings in rats and mice subjected to systemic complement activation by the intravenous injection of cobra venom factor, which is a potent activator of the alternative complement pathway [5–7].

There is now an increasing body of experimental evidence to suggest that hydroxyl radical ($^{\cdot}$OH), which is an iron-mediated conversion product of hydrogen peroxide (H_2O_2), may be the most important oxidant involved in neutrophil-mediated tissue injury [4,6,8,9]. The mechanism(s) of cell and tissue damage produced by oxygen radicals such as $^{\cdot}$OH is only poorly understood. However, it is well accepted that $^{\cdot}$OH which is a very short-lived but highly reactive oxidant can cause lipid peroxidation [10]. In the following communication we will briefly review recently obtained experimental data which demonstrate the appearance in plasma and tissues of lipid peroxidation products following thermal injury of skin.

Lipid Peroxidation Products in Plasma and Tissues

Assays for measurements of products of lipid peroxidation are based on chloroform-methanol extractions of tissues, cells and plasma, allowing for estimation of conjugated dienes, malondialdehyde, hydroperoxides, and fluorochromic products with features of Schiff bases. Procedural details can be found elsewhere [11]. A possible scheme of events in lipid peroxidation of a polyunsaturated fatty acid (PUFA) is depicted in figure 1.

Because neutrophil-derived $^{\cdot}$OH has been implicated in lung microvascular injury following experimental skin burns [4] and because oxidation of polyunsaturated fatty acids by $^{\cdot}$OH and/or iron has been postulated as a mechanism responsible for damage of biological membranes [12], we have examined samples of plasma and tissues from burned and control animals for the presence of products of lipid peroxidation (conjugated dienes) [4]. Our findings are summarized in figure 2. As can be seen, conjugated dienes appear sequentially in both the burned skin (peaking at 15 min postburn) and in the lung (at 60–120 min) as well as in the plasma (with peaks at 30 and 180 min). Extracts from liver, kidney and spleen from thermally injured rats have not shown significant changes in tissue levels of conjugated dienes during the 4-hour observation period (data not shown). It is tempting to speculate that the two

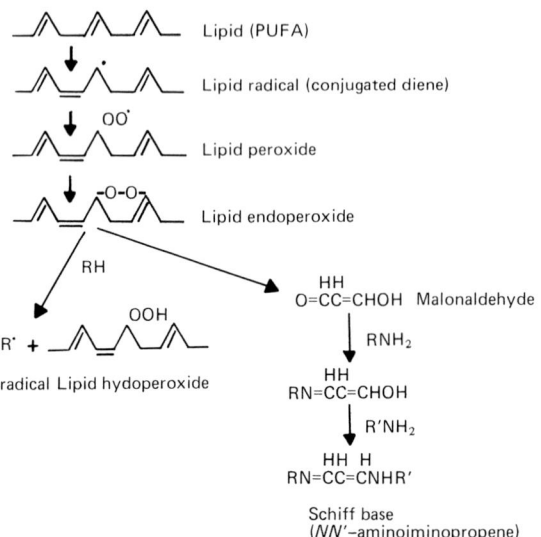

Fig. 1. Postulated steps in lipid peroxidation of a polyunsaturated fatty acid (PUFA). In the initial step, abstraction of hydrogen results in a conjugated diene (lipid radical). This is followed by the formation of a lipid peroxide and its conversion to a lipid endoperoxide. The latter product may then be converted to lipid hydroperoxide or malonaldehyde. Malonaldehyde is known to be highly reactive with epsilon-amino groups of proteins resulting in the appearance of conjugated Schiff bases (fluorochromes).

peaks in plasma of conjugated dienes (at 30 and 180 min) may be connected with the preceeding peaks of conjugated dienes in burned skin and lung tissue at 15 and 120 min, respectively. However, experimental proof of this assumption remains to be secured.

Similar observations of lipid peroxides in thermal injury have previously been reported by *Nishigaki* et al. [13] who have found increases in serum of malondialdehyde (MDA) within 3 h after skin burns (boiling water) in rats. However, in our hands, the thiobarbituric acid assay, which is the most frequently used method for determining lipid peroxidation products (MDA) in vitro, has consistently failed to show significant amounts of MDA in plasma and tissue samples of thermally injured rats. This may in part be explained by the rapid metabolization of MDA in vivo and its inactivation by tissue aldehyde oxidases [14,15]. Furthermore, systemic complement activation in rats induced by cobra venom factor (which results in neutrophil oxidant-dependent acute lung injury)

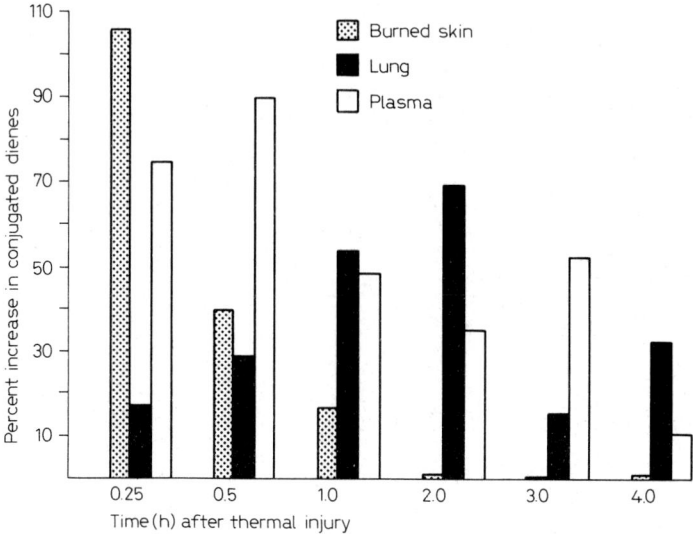

Fig. 2. Kinetics of conjugated dienes in burned skin, plasma and lungs of thermally injured rats.

has also been shown to lack evidence for presence of MDA in plasma despite the fact that increased levels of conjugated dienes, hydroperoxides and fluorochromic substances could clearly be demonstrated [16].

Protection from Lipid Peroxidation

If neutrophil-derived oxygen radicals such as hydroxyl radical are involved in the production of lipid peroxidation products, then interventional measures that result in scavenging of the hydroxyl radical or prevent its generation (by chelation of free iron) should prevent the appearance of conjugated dienes in plasma of thermally injured rats. As shown in figure 3, pretreatment of thermally injured rats with the hydroxyl radical scavenger dimethyl sulfoxide (DMSO) significantly reduces (by 68%) the 3-hour peak in plasma levels of conjugated dienes. A similar degree of protection from lipid peroxidation can also be achieved by pretreatment of burned animals with the iron chelator deferoxamine mesylate. As would be expected, iron-saturated deferoxamine exhibits no protective effects. Quite to the contrary, the iron-saturated chelator appears to

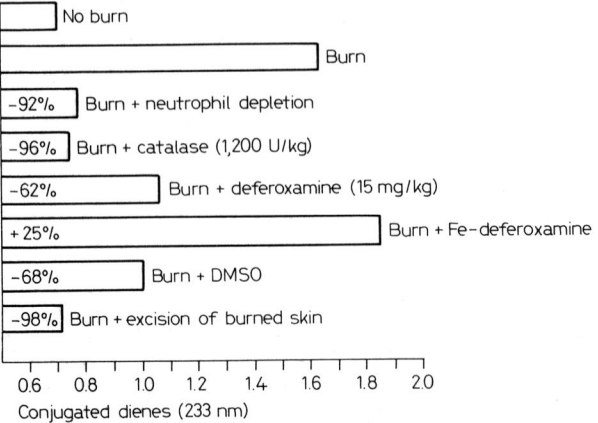

Fig. 3. Effect of interventional measures in thermally injured rats on 3-hour plasma levels of conjugated dienes.

slighthly augment the generation of conjugated dienes, presumably by providing iron ions to neutrophils.

The role of iron in the in vivo production of lipid peroxides is not completely understood but may be explained by its essential role as a redox agent in the conversion of hydrogen peroxide to the highly toxic hydroxyl radical (Fenton reaction). This also explains the potent protection from lipid peroxidation afforded by treatment of animals with catalase (fig. 3), which converts hydrogen peroxide to dioxygen and water, thus preventing the iron-catalyzed production of hydroxyl radical from hydrogen peroxide. When the major source of oxygen-derived free radicals is abolished as demonstrated in neutrophil-depleted rats, a profound protection from lipid peroxidation in thermally injured rats is also observed (fig. 3). Accordingly, a strong relationship between the numbers of available blood neutrophils and the plasma content of conjugated dienes has been demonstrated 3 h after thermal injury [4]. Protection from in vivo lipid peroxidation by interventional measures that result in scavenging of hydroxyl radical (DMSO) or prevent its production (neutrophil depletion, catalase, iron chelation) has also been observed in rats undergoing acute microvascular injury in the lung following intravenous infusion of the complement activator, cobra venom factor [16].

The most striking reduction in the appearance of plasma conjugated dienes can be seen in thermally injured rats by an immediate excision

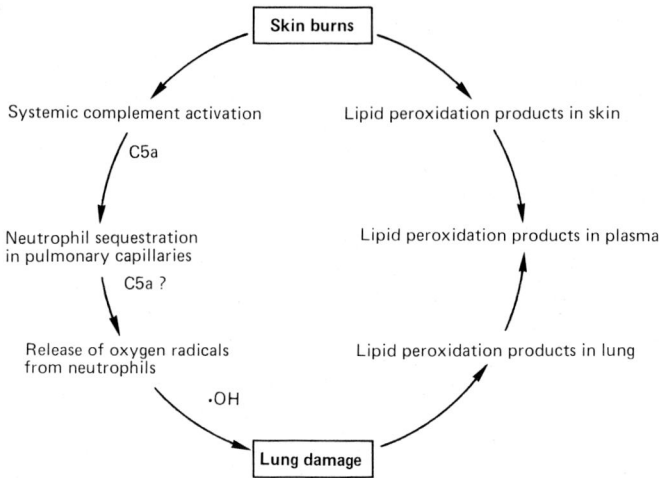

Fig. 4. Postulated scheme of pathophysiologic events leading to acute lung injury and appearance of lipid peroxidation products secondary to skin burns.

(within 3 min) of the skin that has been thermally injured (fig. 3). This manipulation not only completely protects against lipid peroxidation but also prevents systemic complement activation and the development of acute lung injury [4]. It appears likely that one of the first events in the thermally injured animal is the activation of the complement system in the vascular bed of the burned skin. This may stimulate blood neutrophils in the burn area to release oxygen radicals which, in turn, may cause the evidence of lipid peroxidation as revealed by the appearance in burned skin and plasma of conjugated dienes which achieve peak levels at 15 and 30 min postburn, respectively (fig. 2). These rather localized early events may with time develop into a generalized process of systemic complement activation, which ultimately leads to neutrophil-dependent and oxygen radical-mediated acute lung injury. As a result of these lung pathogenic events secondary to thermal trauma of skin, lipid peroxidation products appear in lung tissue and plasma at 120 and 180 min, respectively. A hypothetical scheme of these events is depicted in figure 4.

The origin and the chemical nature of lipid peroxidation products (conjugated dienes) appearing in tissues and plasma of thermally injured rats are not yet known. It is also not known whether these lipid peroxides per se cause tissue damage or merely represent byproducts of the oxygen radical-induced tissue damage. It should be noted that the intravenous injection of catalase 45 min after thermal injury, at a time point at which the first peak of conjugated dienes has already appeared in plasma (fig. 2), completely prevents lung injury and the appearance in plasma of the 3-hour peak of conjugated dienes [4]. In other words, plasma conjugated dienes found at 30 min after thermal injury are not themselves sufficient to cause lung injury. This appears to support the concept that conjugated dienes are secondary products of oxygen radical-mediated tissue damage. On the other hand, based on studies by *Nishigaki* et al. [13], it has been suggested that lipid peroxidation products entering the bloodstream at the site of the skin burn may within 2–20 days thereafter bring about damage to cells of remote other organs.

Conclusion

Thermal injury to skin of rats results in acute lung injury and appearance of products of lipid peroxidation, both of which can be prevented by immediate excision of the burned skin or by treatment of experimental animals with catalase, scavengers of hydroxyl radical, or iron chelators. Lipid peroxidation products (conjugated dienes) appear in the burned skin, lungs, and plasma. The derivation of the plasma-conjugated dienes in thermally injured rats appears to be related to complement-activated blood neutrophils and their production of hydrogen peroxide and hydroxyl radical. The origin and chemical nature of these lipid peroxidation products are not yet known, but it appears likely that the conjugated dienes may originate from cell membranes of damaged vascular endothelial cells in burned skin and lung and/or may constitute oxidized lipids and lipoproteins of plasma and blood cell membranes. These findings may provide some suggestions for therapeutic interventions in burned patients and perhaps other clinical situations where oxygen radicals may be involved in cell and tissue injury.

References

1   Gelfand, J. A.; Donelan, M.; Burke, J. F.: Preferential activation and depletion of the alternative complement pathway by burn injury. Ann. Surg. *198:* 58–62 (1983).
2   Till, G. O.; Beauchamp, C.; Menapace, D.; Tourtellotte, W. W.; Kunkel, R.; Johnson, K. J.; Ward, P. A.: Oxygen radical dependent lung damage following thermal injury of rat skin. J. Trauma *23:* 269–277 (1983).

3 Heideman, M.: Complement activation in vitro induced by endotoxin and injured tissue. J. surg. Res. *26:* 670–673 (1979).

4 Till, G. O.; Hatherill, J. R.; Tourtellotte, W. W.; Lutz, M. J.; Ward, P. A.: Lipid peroxidation and acute lung injury after thermal trauma to skin. Evidence of a role for hydroxyl radical. Am. J. Path. *119:* 376–384 (1985).

5 Till, G. O.; Johnson, K. J.; Kunkel, R.; Ward, P. A.: Intravascular activation of complement and acute lung injury. Dependency on neutrophils and toxic oxygen metabolites. J. clin. Invest. *69:* 1126–1135 (1982).

6 Ward, P. A.; Till, G. O.; Kunkel, R.; Beauchamp, C.: Evidence of role of hydroxyl radical in complement and neutrophil-dependent tissue injury. J. clin. Invest. *72:* 789–801 (1983).

7 Tvedten, H. W.; Till, G. O.; Ward, P. A.: Mediators of lung injury in mice following systemic activation of complement. Am. J. Path. *119:* 92–100 (1985).

8 Fligiel, S. E. G.; Ward, P. A.; Johnson, K. J.; Till, G. O.: Evidence for a role of hydroxyl radical in immune-complex-induced vasculitis. Am. J. Path. *115:* 375–382 (1984).

9 Fox, R. B.: Prevention of granulocyte-mediated oxidant lung injury in rats by a hydroxyl radical scavenger, dimethyl thiourea. J. clin. Invest. *74:* 1456–1464 (1984).

10 Fong, K.-L.; McCoy, P. B.; Poyer, J. L.: Evidence that peroxidation of lysosomal membranes is initiated by hydroxyl free radicals produced during flavin enzyme activity. J. biol. Chem. *248:* 7792–7797 (1973).

11 Buege, J. A.; Aust, S. O.: Microsomal lipid peroxidation; in Fleischer, Packer, Methods in enzymology, vol. 52, pp. 302–310 (Academic Press, New York 1978).

12 Barber, A. A; Bernheim, F.: Lipid peroxidation: its measurement, occurrence, and significance in animal tissues. Adv. gerontol. Res. *2:* 355–401 (1967).

13 Nishigaki, J.; Hagihara, M.; Hiramatsu, M.; Izawa, Y.; Yagi, K.: Effect of thermal injury on lipid peroxidase levels of rat. Biochem. Med. *24.:* 185–189 (1980).

14 Placer, Z.; Veselkova, A.; Rath, R.: Kinetik des Malondialdehyds im Organismus. Experientia *21:* 19–20 (1965).

15 Horton, A. A.; Packer, L.: Mitochondrial metabolism of aldehydes. Biochem. J. *116:* 16P (1970).

16 Ward, P. A.; Till, G. O.; Hatherill, J. R.; Annesley, T. M.; Kunkel, R.: Systemic complement activation, lung injury, and products of lipid peroxidation. J. clin. Invest. *76:* 517–527 (1985).

G. O. Till, MD, Department of Pathology, University of Michigan Medical School, Ann Arbor, MI 48109 (USA)

Novelli, Ursini (eds.), Oxygen Free Radicals in Shock. Int. Workshop,
Florence 1985, pp. 42–45 (Karger, Basel 1986)

HPLC Analysis of Tissue Thiobarbituric Acid-Reactive Material during Endotoxic Shock in Rats

M. Maiorino[a], L. Bordi[b], G. Consales[b], F. Ursini[a], G. P. Novelli[b]

[a] Institute of Biological Chemistry, University of Padova;
[b] Institute of Anesthesiology, University of Florence, Italy

Enhanced free radical generation leading to lipid peroxidation has been claimed to be associated with various disorders [1]. Furthermore, it was recently claimed that endotoxic shock is associated with an increased lipid peroxidation rate and that lipid peroxidation could be the molecular mechanism of tissue and microvascular injury [2, 3]. However, in vivo the quantitative evaluation of lipid peroxidation is a difficult task, mainly due to the absence of a stable lipid peroxidation product to be measured. The lipid hypoperoxides indeed, the initial and major lipid peroxidation products are not stable due to the chemical reactivity [4] and the enzymatic reduction [5, 6]. Therefore, the evaluation of lipid peroxidation in vivo is generally carried out by the analysis of the lipid hydroperoxide breakdown products: volatile hydrocarbons, hydroxyaldehydes, malondialdehyde (MDA). The most simple and sensitive procedure is the spectrophotometric determination of MDA or MDA precursors, measured as thiobarbituric acid-reactive material (TBA-RM). This analysis is highly sensitive but it presents some disadvantages: only a small percentage of hydroperoxides are transformed in MDA, MDA can be oxidized in the mitochondria and thiobarbituric acid reacts with other substances present in tissue homogenates. On the other hand, since the extent of lipid peroxidation in tissues in vivo under different experimental pathological conditions is very low, only the TBA test seems to offer the required sensitivity. In order to increase the specificity and possibly the sensitivity of this test, we used a high-performance liquid chromatography (HPLC) analysis of the chromogen produced by the interaction of thiobarbituric acid and MDA. This analysis allowed a careful evaluation of the lipid peroxidation in a number of rat organs following the induction of an endotoxic shock by the i.p. administration of *E. coli* endotoxin.

Materials and Methods

Male Wistar rats (250 g), after an overnight fasting, were treated with *E. coli* lipopolysaccharide 0111:B4 (40 mg/kg i.p. in saline). A second group of rats was treated with aspirin (20 mg/kg i.p.) 15 h before the experiment, a third group was treated twice with vitamin E (500 mg/kg in olive oil i.p.) 48 and 24 h before the experiment. The animals were killed by cervical dislocation, the organs removed, and homogenized (1:4 w:v) using a Polytron homogenizer in 0.1 M Tris-HCl 0.3 M KCl buffer at pH 7.5, containing 0.01% butylabed hydroxytoluene (BHT), to avoid further peroxidation. To 0.5 ml of homogenate 1 ml of 20% TCA containing 0.01% BHT was added. The mixture was then boiled for 10 min in glass-stoppered tubes. After cooling and centrifugation for 15 min at 3,000 g, 1 ml of supernatant was withdrawn and added to 2 ml of 0.25% thiobarbituric acid in Tris glycine 0.2 M, pH 3, containing 0.01% BHT. After boiling for 30 min in glass-stoppered tubes, the samples were filtered and injected into the HPLC. The column used was an Ultrasphere ODS (Beckman Co.), the mobile phase 8% acetonitrile, 15% terahydrofurane, 10 mM esansulfonic acid, 10 mM orthophosphoric acid, pH 3; the flow rate 0.7 ml/min. Malondialdehyde-tetraethylacetal was used as standard. All reagents were of analytical grade, the HPLC solvents was filtered and degassed before the use.

Results and Discussion

HPLC procedures have been recently described for the analysis of MDA in biological samples either as free aldehyde [7] or as TBA complex [8]. The latter procedure seems more useful for the in vivo investigations where the specific chromogen is generated from the reaction between bicyclic endoperoxides and TBA [9] and where MDA, when produced, is rapidly oxidized by the mitochondria. The procedure described by *Bird* et al. [8] was modified in order to increase the resolution of the specific peak in samples containing 535-nm absorbing contaminants. The esansulfonic acid acting as a counter ion increased the affinity of the chromogen for the stationary phase and allowed wider modifications of the polarity of the mobile phase useful for the resolution of the peak of the specific complex in the presence of contaminants.

The results of the determination of TBA-RM in rat tissues 120 min following the endotoxin administration are reported in table I. The increase was significant in the heart, liver and lung while in the kidney and

Table I. Thiobarbituric acid-reactive material (TBA-RM) in rat tissues following *E. coli* endotoxin treatment: effect of aspirin and vitamin E

Organ	TBA-RM, pmol/mg of protein			
	control	endotoxin	endotoxin-aspirin	endotoxin-vitamin E
Heart	105.4 (20.3)	155.6 (28.6)	133.5 (25.3)	110.3 (22.2)
Kidney	136.5 (18.6)	131.3 (19.5)	134.2 (18.3)	110.1 (15.5)
Liver	86.2 (9.3)	95.9 (9.6)	95.9 (8.4)	71.8 (9.5)
Lung	83.5 (7.8)	95.5 (9.1)	96.7 (8.4)	81.6 (8.9)
Brain	264.8 (19.6)	283.6 (23.4)	285.4 (26.7)	258.2 (28.6)

The animals were treated and TBA-RM was measured as described in the text. The values represent the mean between four different determinations and the standard errors.

brain no appreciable variations were observed. The time course of the increase showed marked individual variations and the maximum reached within 2 h was followed by a slow decrease (not shown). The TBA-RM generated under these conditions is not a by-product of prostaglandin biosynthesis because it is not prevented by the aspirin treatment (to inhibit cyclooxigenase activity). On the other hand, vitamin E completely prevented the TBA-RM production supporting the hypothesis that a higher lipid peroxidation rate takes place during endotoxic shock.

In conclusion, the present data, obtained by an analytic approach, much more sensitive and specific than the usual spectrophotometric TBA test, confirm the previous report that lipid peroxidation occurs during endotoxemia. On the other hand, it is the opinion of the authors that further efforts are still needed to identify some other more stable compounds generated during lipid peroxidation, to evaluate quantitatively the peroxidation rate in vivo.

References

1   Bulkey, G. B.: The role of oxygen free radicals in human disease processes. Surgery, St Louis *94:* 407–411 (1983).
2   Yoshikawa, T.; Murakami, M.; Furukawa, Y.; Kato, H.; Takemura, S.; Kondo, M.: Lipid peroxidation and experimental disseminated intravascular coagulation in rats enduced by endotoxin. Thromb. Haemostasis *49:* 214–216 (1983).

3   Ogawa, R.; Morita, T.; Kunimoto, F.; Fujita, T.: Changes in hepatic lipoperoxide concentration in endotoxemic rats. Circul. Shock 9: 369–374 (1982).

4   Pryor, W. A.; Castle, L.: Chemical methods for the detection of lipid hydroperoxides; in Packer, Methods in enzymology. Oxygen radicals in biological systems, No. 105, pp. 293–299 (Academic Press, New York 1984).

5   Maiorino, M.; Roveri, A.; Ursini, F.; Gregolin, C.: Enzymatic determination of membrane lipid peroxidation. J. Free Radic. Biol. Med. (in press).

6   Ursini, F.; Maiorino, M.; Gregolin, C.: The selenoenzyme phospholipid hydroperoxide glutathione peroxidase. Biochim. biophys. Acta 839: 62–70 (1985).

7   Lang, J.; Hecknast, P.; Esterbauer, H.; Slater, T. F.: Detection of malondialdehyde by HPLC; in Bors, Saran, Tait, Oxygen radicals in chemistry and biology, pp. 351–354 (de Gruyter, Berlin 1984).

8   Bird, R. P.; Hung, S. O. O.; Hadley, M.; Draper, H. H.: Determination of malonaldehyde in biological materials by high pressure liquid chromatography. Analyt. Biochem. 128: 240–244 (1983).

9   Pryor, W. A.; Stanley, J. P.: A suggested mechanism for the production of malonaldehyde during the autoxidation of polyunsaturated fatty acid. Non-enzymatic production of prostaglandin endoperoxides during autoxidation. J. org. Chem. 40: 3615–3617 (1975).

M. Maiorino, MD, Institute of Biological Chemistry, University of Padova, Via Marzolo 3, I-35100 Padova (Italy)

Novelli, Ursini (eds.), Oxygen Free Radicals in Shock. Int. Workshop,
Florence 1985, pp. 46–52 (Karger, Basel 1986)

Fluidity of Mitochondrial and Microsomal Membranes in Traumatic Shock as Studied by ESR

*Gian Paolo Novelli[a], Giacomo Martini[b], Lorenzo Bordi[a],
Guglielmo Consales[a], Sauro Porciani[c], Aldo Becciolini[c]*

[a]Institute of Anaesthesiology and Intensive Care, University of Florence;
[b]Department of Chemistry, University of Florence; [c]Laboratory of Radiobiology,
Department of Clinical Physiopathology, University of Florence, Italy

Introduction

The electron spin resonance spectroscopy (ESR) of spin labels introduced into biological membranes is a valuable tool for a direct determination of the changes in the membrane's physicochemical status brought about by several processes [1, 2] including peroxidation [3, 4].

Shock seems to be provoked by an overflow of oxygen radicals and peroxidative damage is therefore expected [*Ogawa* et al., this meeting].

This would be consistent in a decrease of membrane fluidity and a change of the molecular order, both causative of impairment of several functions like selective permeability, enzyme activity, ion transport, etc. [5].

To search for and characterize the membrane's damage related to oxygen-derived free radicals, the physicochemical status of mitochondrial and microsomal membranes, isolated from lethally shocked rats, was studied by means of ESR of two spin labels, entering in different positions inside the lipid bilayers.

Materials and Methods

Preparation of Membranes. Experiments were performed on male Wistar rats $(200 \pm 10$ g) fasted 12 h before experiments.

Shock was provoked by a rotating drum submitting animals to 1,200 revolutions, a dose known to be 100% lethal.

Immediately after rotation, rats were killed by cervical dislocation and the liver was excised; the same was made on normal rats.

The membranes were prepared as described by *Mugnai and Boddi* [6].

The livers were immediately kept in a 0.25-*M* cold sucrose solution buffered with tris/HCl 10 m*M* (pH 7.5) and EDTA (sodium salt) 0.1 *M*. Then the livers were weighed and homogenized with an UltraTurrax homogenizer in 4 volumes of the same solution.

Nuclei and cell fragments were sedimented after centrifugation at 600 *g* for 10 min. Mitochondria were separated at 5,000 *g* for 10 min in a refrigerated centrifuge.

The 'fluffy layer' of the sediment was poured off leaving behind the well-packed mitochondria.

To prepare the pure mitochondrial fraction, the pellet was resuspended and washed 4–5 times in the same solution and centrifuged at 5,000 *g* for 10 min.

The supernatant was centrifuged at 20,000 *g* for 15 min and the sediment was discarded; a further centrifugation at 104,000 *g* for 60 min was used to obtain the microsomal fraction. Protein content was determined by the usual Lowry method.

Spin Labeling. Two different nitroxide spin labels were used, whose paramagnetic moieties localize near the surface and in the depth of membrane, respectively. The nitroxides were derivates of stearic acid:

$$CH_3-(CH_2)_m-\overset{\displaystyle}{\underset{\displaystyle O\quad N-O}{C}}-(CH_2)_n-COOH$$

with m=12, n=3 (5-SASL) and m=1, n=14 (16-SASL) were purchased by Syva Ass. (Palo Alto, Calif.) and used without further purification. Stock $2.5 \cdot 10^{-3}$ mol/l solutions were prepared in anhydrous ethanol. The appropriate amount of the SASL solution to get a final membrane protein/SASL ratio 250:1 as used in each experiment and placed at the bottom of a test tube. After solvent removal with an oxygen-free nitrogen stream, the spin labels were added to 50 mg of the isolated membranes (resuspended in 0.5 ml saline and vigorously stirred for at least 5 min).

ESR Spectra. ESR spectra were taken with a Bruker 200 D spectrometer by using either flat cell for liquid samples or quartz capillary, 1 mm i.d. In the latter case temperature was controlled with the Bruker ST 100/700 variable temperature accessory and kept at 25 °C.

Instrumental settings (gain, modulation, amplitude, power) were such to avoid signal overmodulation and saturation.

From the ESR spectra the following parameters relevant for the motion were evaluated: (a) the correlation time for the reorientational motion of the $>$N-O group, τ_c; (b) the order parameter S; (c) the wobbling parameter γ.

The correlation times for the motion were evaluated by computing the ESR spectra with the program written by *Freed* [7] and modified as by *Martini* [8]. The relevant ESR parameters used in the simulation were:

$$
\begin{array}{ll}
g_{xx} = 2.0088 & A_{xx} = 6.3 \text{ G} \\
g_{yy} = 2.0061 & A_{yy} = 5.8 \text{ G} \\
g_{zz} = 2.0027 & A_{zz} = 33.6 \text{ G}
\end{array}
$$

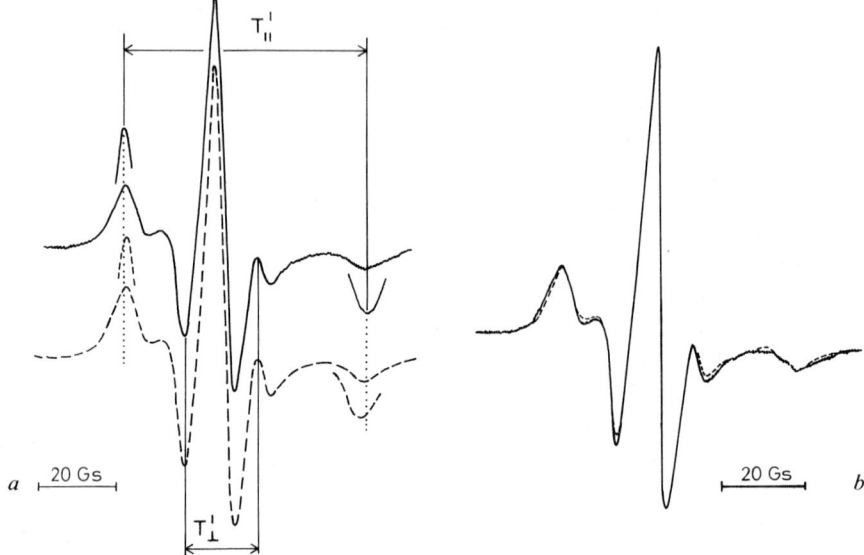

Fig. 1. a ESR spectra at room temperature of 5-SASL introduced in mitochondrial membranes of shocked rats (full line) and in control rats (dashed line). $T_{\parallel}$ and $T_\perp$ were used for the calculation of order parameters S and γ. *b* Comparison between experimental (full line) and computed spectra (dotted line) of 5-SASL in mitochondrial membranes.

The S parameter is a measure of the average amplitude of membrane lipid acyl chain motion. This parameter is widely used in membrane studies and several expressions are given [1, 9, 10]. We used the simplified model given by *Gaffney and Lin* [11] for fatty acid spin labels:

$$S = \frac{T_{\parallel}' - (T_\perp' + C)}{T_{\parallel}' + 2(T_\perp' + C)} \times 1.723, \quad C = 1.4 - 0.053\,(T_{\parallel}' - T_\perp'),$$

where the meaning of $T_{\parallel}'$ and $T_\perp'$ is shown in figure 1. The term C is a correction to the experimentally determined $T_{\parallel}'$ and $T_\perp'$, based on the calculated spectra. Its limits are S=1 for a static situation and S=0 for very fast motion with complete averaging of the magnetic anisotropy.

The wobbling parameter γ refers to the fluctuations of the angle ϑ_1 between the spin label molecular axis R and the axis Z normal to the membrane: ϑ_1 may indeed randomly fluctuate between $\vartheta_1 = 0$ and the maximum half amplitude $\vartheta_1 = \gamma$.

When $\gamma = 0$, the wobble model reduces to rotation about R, that means rigid conditions; contrarily, when the amplitude motion is at its largest value $\gamma = \pi/2$, the whole anisotropy is removed and the spin label freely tumbles as in a non-viscous liquid solution. The γ parameter was calculated following the procedure reported in [11].

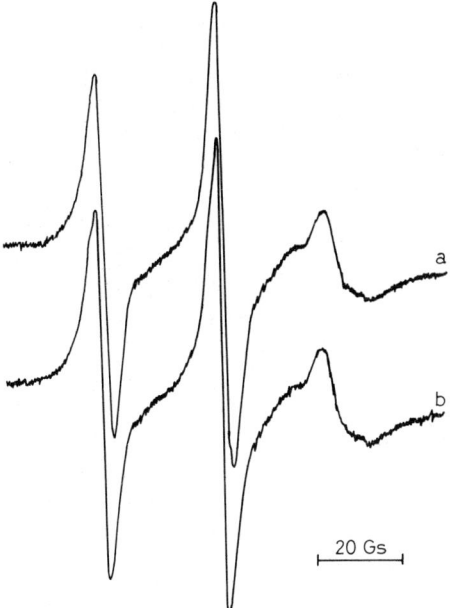

Fig. 2. ESR spectra of 16-SASL in mitochondrial membranes of shocked rats (a) and of control rats (b).

Results

Figure 1a shows a typical ESR spectrum of 5-SASL in membranes of liver mitochondria of rats submitted to traumatic shock. As a comparison the ESR spectrum of control rat is also shown. Both spectra were typical of slow-moving radicals in which the anisotropy of g and A tensors were not averaged [12, 13].

The fit between experimental and calculated spectra is shown in figure 1b. Correlation time $\tau_{,,}$ and $\tau_{\perp}$ of $6.9 \cdot 10^{-9}$s and $1.8 \cdot 10^{-8}$s for mitochondria of shocked rats and $6.7 \cdot 10^{-9}$s and $1 \cdot 10^{-8}$s for control ones were valued respectively. Almost the same line shapes with the same correlation times were obtained for the microsomial fraction. Because the motion of the radical was anisotropic, two different correlation times were used in the calculation of the tumbling around Z axis ($\tau_{,,}$) and around χ and υ axes ($\tau_{\perp}$) of the paramagnet.

Figure 2 shows the ESR spectra of 16-SASL in the mitochondrial fraction of liver of shocked and control rats. Differently as with 5-SASL,

Table I. Correlation times for the motion τ and membrane order parameters S and γ of stearic acid nitroxides in rat liver mitochondria and microsomes[1]

System	$\tau_{\parallel}$ $\tau_{\perp}$ ($\times 10^{-9}$s)		S	γ degree	[A], Gs
5-SASL-Mitochondria, shocked (11)	6.9	10	0.711±0.008	34±1	15.29±0.06
5-SASL-Mitochondria, control (5)	6.7	10	0.675±0.006	37±1	15.21±0.06
5-SASL-Microsomes, shocked (9)	6.9	10	0.714±0.013	33±1	15.29±0.06
5-SASL-Microsomes, control (5)	6.7	10	0.678±0.06	36±1	15.26±0.07
16-SASL-Mitochondria, shocked (9)	2.4	4.0	0.13 ±0.02	76±2	13.70±0.07
16-SASL-Mitochondria, control (4)	2.4	4.0	0.12 ±0.02	78±2	13.74±0.06
16-SASL-Microsomes, shocked (6)	2.4	4.0	0.13 ±0.02	77±2	13.68±0.05
16-SASL-Microsomes, control (3)	2.4	4.0	0.12 ±0.02	78±2	13.65±0.08

[1] Data given as mean ± SD. Values in parentheses are number of experiments.

no appreciable differences were observed in the ESR line shape and in the magnetic parameters.

The correlation times also did not differ in shocked and control samples. The calculated values ($\tau_{\parallel}=2.4 \cdot 10^{-9}$s; $\tau_{\perp}=4.0 \cdot 10^{-9}$s) indicated an intermediate situation between fast and slow motion conditions.

Table I reports the values of τ, S and γ as calculated as reported above. The values of these parameters showed that for the 16-SASL, that was positioned deeper in the membrane, larger segmental motion occurred than in the region closer to the polar-non-polar interface where 5-SASL was located. However, the data shown in table I indicated that when 5-SASL was used S and γ-values in shocked rats were higher and lower respectively than in control rats. No significant differences were observed in S and γ of 16-SASL.

As with order parameters, the hyperfine coupling constant [A] also showed a dependence on the position that gives the so-called polarity profile through the membrane [14]. From the data shown in table I, no differences existed in polarity profile of shocked and control rats either in mitochondrial or in microsomal membranes.

Discussion

Physicochemical damage of cell membranes appeared to be similar in both mitochondria and microsomes but to be different in the various membrane layers, polarity and fluidity being unmodified as the paramagnetic moiety of the spin-label was located deeper in the bilayer. In other words, results were indicative of a lessening of peroxidative damage from the hydrophilic surface towards the hydrophobic core of the membrane bilayer.

This means that membrane damage consequent to shock occurred mostly near the membrane surface, without appreciable changes in the polarity profile.

Similar dependence of the fluidity changes on the membrane depth were also observed after peroxidation of rat microsomal membrane phospholipids [3] and of sonicated soybean phospholipid vesicles [4].

The action of O_2^-, $OH^·$, O_2^*, etc. on unsaturated fatty acid in membranes results in formation of peroxy radicals, endoperoxides and hydroperoxides. The final products should contain oxygen-linked bridges between adjacent acyl chains resultant in physical changes in the molecular order of the hydrocarbon chains of membrane phospholipids.

It has been suggested by *Nagy* et al. [15] that the free radical induced damage of membranes results in an increased cross-linking of the components. This leads to a decreased permeability to ions and water with resultant impairment of the membrane function.

The distribution of lessened fluidity in cell membranes from shocked rats seems to be in agreement with lipid peroxidation induced by oxygen free radicals attacking membranes from the outer towards the inner layers.

In addition, peroxidated lipid chains, being more hydrophilic, turn their axis towards the outer layer with a resultant decrease of fluidity.

According to the results reported here, it is confirmed that during shock, the oxygen radicals act on the cell membranes provoking a loss of double bonds in the unsaturated fatty acids located mainly near the polar group. This distribution of damage agrees (a) with the hydrophilic nature of oxygen derived free radicals that prevents a deep penetration inside the membrane bilayer; (b) with a disposal of peroxidated chains near to the outer layers of the membrane bilayer. The peroxidative damage so demonstrated is the cause of the membrane damages typical of shock states.

References

1 Marsh, D.: Electron spin resonance: spin labels; in Grell, Membrane spectroscopy, pp. 51–142 (Springer, New York 1981).

2 Fung, L. W. N.; Johnson, M. E.: Recent developments in spin label EPR methodology for biomembrane spectroscopy; in Lee, Current topics in bioenergetics, vol. 13, pp. 107–157 (Academic Press, New York 1984).

3 Bruch, R. C.; Thayer, W. S.: Differential effect of lipid peroxidation on membrane fluidity as determined by electron spin resonance probes. Biochim. biophys. Acta 733: 216–222 (1983).

4 Curtis, M. T.; Gilfor, D.; Farber, J. L.: Lipid peroxidation increases the molecular order of microsomal membranes. Archs Biochem. Biophys. 235: 644–649 (1984).

5 Wendel, A.; Feuerstein, S.: Drug-induced lipid peroxidation in mice. Modulation by monooxygenase activity, glutathione and selenium status. Biochem. Pharm. 30: 2513–2520 (1981).

6 Mugnai, G.; Boddi, V.: Stearoyl-Coa desaturase in mitochondrial membrane fractions. It. J. Biochem. 26: 245–253 (1977).

7 Freed, J. H.: Theory of slow tumbling ESR spectra for nitroxides; in Berliner, Spin labeling. Theory and applications, vol. I, pp. 53–132 (Academic Press, New York 1976).

8 Martini, G.: ESR study of a negative spin probe adsorbed from a silica/water dispersion. Colloids Surfaces 11: 409–421 (1984).

9 Seelig, J.: Anisotropic motion in liquid crystalline structures; in Berliner, Spin labeling. Theory and applications, vol. I, pp. 373–408 (Academic Press, New York 1976).

10 Schreier, S.; Polnaszek, C. F.; Smith, I. C. P.: Spin labels in membranes. Problems in practice. Biochim. biophys. Acta 515: 395–436 (1978).

11 Gaffney, B. J.; Lin, D. C.: Spin label measurements of membrane-bound enzymes; in Martonosi, The enzymes of biological membranes, vol. I, pp. 71–90 (Plenum Publishing, New York 1976).

12 Goldman, S. A.; Bruno, G. V.; Freed, J. H.: Estimating slow motional correlation times for nitroxides by electron spin resonance. J. phys. Chem. 76: 1858–1860 (1972).

13 Romanelli, M.; Ottaviani, M. F.; Martini, G.: ESR study on the interaction between a positively changed spin probe and the silica gel surface. J. Colloid Interface Sci. 96: 373–380 (1983).

14 Harris, J.; Power, T. J.; Bieber, A. L.; Watts, A.: An ESR spin label study of the interaction of purified mojave toxin with synaptosomal membranes from rat brain. Eur. J. Biochem. 131: 559–565 (1983).

15 Nagy, K.; Simon, P.; Nagy, I. Zs.: Spin label studies of synaptosomal membranes of rat brain cortex during aging. Biochem. biophys. Res. Commun. 117: 688–694 (1983).

Prof. G. P. Novelli, Institute of Anaesthesiology and Intensive Care, University of Florence, Policlinico di Careggi, V. le Morgagni, I-50134 Florence (Italy)

Novelli, Ursini (eds.), Oxygen Free Radicals in Shock. Int. Workshop,
Florence 1985, pp. 53–55 (Karger, Basel 1986)

The Influence of N-t-Butyl-α-Phenylnitrone on the Stability of Erythrocyte Membranes

R. I. Petrova, G. P. Novelli, S. M. Iovtchev, M. Z. Metodieva,
N. A. Nicolov

Medical Academy, Medico-Biological Institute, Sofia, Bulgaria;
Institute of Anesthesiology and Intensive Care, Florence, Italy

During endotoxin shock the erythrocytes and their membranes change their mechanical properties, such as deformability and stability [1, 2]. It is known that during shock in the lipid bilayer of membranes, free radicals occur which change its properties. Oxygen free radicals in shock change the membrane permeability [3, 4]. Positive results are obtained by the shock treatment with antioxydants and especially with N-t-butyl-α-phenylnitrone (NBP). The aim of the present study is to investigate and compare the influence of endotoxin and NBP on the erythrocyte membranes stability.

Material and Methods

Experiments were carried out in 7 male rabbits (Belgian giant breed with an average weight of 3.5 ± 0.1 kg). Blood specimens for membrane stability measurements were withdrawn from the auricular vein.

Erythrocytes were separated by a 3-fold centrifugation (1,000 g, 10 min) in saline. Erythrocytes (10^9 cells·ml^{-1}) were incubated in an endotoxin-saline solution ($2.5 \cdot 10^{-2}$ and $2.5 \cdot 10^{-4}$ mg·ml^{-1}) for 2 h at 37 °C. *E. coli* ($O_{III}:B_4$) endotoxin was used.

Rabbit erythrocytes (10^9 cells·ml^{-1}) were incubated in an NBP-saline solution (1 and 2 mg·ml^{-1}) for the same time and the same temperature. The control sample was erythrocytes (10^9 cells·ml^{-1}) resuspended in saline.

The hemolytic process was estimated with spectral colorimeter 'Specol-10' (GDR). The optical density of suspension was registrated with a potentiometric recorder (Budapest). The hemolytic agent was 0.002 N hydrochloric acid [5].

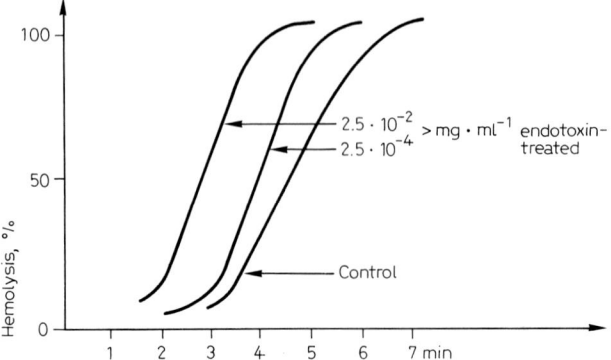

Fig. 1. Changes in stability of erythrocyte membranes treated with two concentrations of endotoxin ($2.5 \cdot 10^{-2}$ and $2.5 \cdot 10^{-4}$ mg·ml⁻¹).

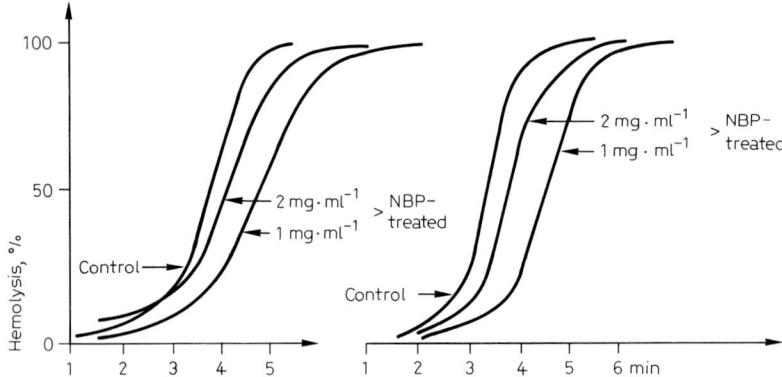

Fig. 2. Changes in stability of erythrocyte membranes of 2 animals treated with 2 concentrations of NBP (1 and 2 mg·ml⁻¹).

Results and Discussion

After incubation with endotoxin-saline solution, erythrocytes hemolyze more rapidly than control samples, i.e. their membrane stability decreases. Figure 1 shows the process of hemolysis of red blood cells. The erythrocyte suspensions were treated with $2.5 \cdot 10^{-2}$ mg·ml⁻¹ endotoxin. The process of hemolysis was accelerated as compared to the control samples. By incubation with lower concentration endotoxin (10^{-4} mg·ml⁻¹) this change decreased (fig. 1).

The process of hemolysis of erythrocyte suspensions treated with antioxidant (NBP) is shown in figure 2. N-t-Butyl-α-phenylnitrone slowed down the hemolysis process, i.e. the cell membranes became more stable. The use of 2-fold lower concentration of NBP increased its stabilizing effect (fig. 2).

Conclusions

(1) *E. coli* endotoxin at concentrations 10^{-4} and 10^{-2} mg·ml^{-1} decreases the stability of red cell membranes. (2) Incubation of erythrocyte suspensions with 1 and 2 mg·ml^{-1} NBP-saline increases their membrane stability, i.e. the process of hemolysis is slower. The lower concentration of NBP has greater stabilizing effect on rabbit red blood cell membranes.

References

1   Iovtchev, S. M.; Stoeff, S. V.; Metodieva, M. Z.; Petrova, R. I.; Stoylov, S. P.: Comparative study of the deformability of RBC membranes, BBBD, Plovdid, BG 1985.
2   Petrova, R. I.; Iovtchev, S. M.; Metodieva, M. Z.; Nicolov, N. A.: Deformability of erythrocytes in endotoxin shock, Symp. Pathologische Biochemie des Schocks, Dresden 1984.
3   Novelli, G. P.; Gaudio, A. R.: Oxygen free-radicals in shock states; in Lewis, Haglund, Shock research, Fernström Fdn Ser., vol. 3, pp. 31–42 (1983).
4   Robbins, S. L.; Cotran, R. S.; Kumar, V.: Inflammation and repair; in Pathologic basis of disease, p. 50 (Saunders, Philadelphia 1984).
5   Terskov, I. A.; Gitelson, I. I.: Method kislotnich erytrogram. Biophysika, Moscow 2: 259 (1957).

Prof. N. A. Nicolov, Department of Pathophysiology, 1, G. Sofiisky Street, Sofia 14–31 (Bulgaria)

Novelli, Ursini (eds.), Oxygen Free Radicals in Shock. Int. Workshop,
Florence 1985, pp. 56–59 (Karger, Basel 1986)

Plasma Mediators of Oxygen Free Radicals in Shock

Mats Heideman

Department of Surgery I, University of Göteborg, Sweden

Introduction

Polymorph nuclear leukocytes, whose resting metabolism is based
largely on anaerobic glycolysis, display an increase in oxygen consumption upon phagocytosis of bacteria or other particles. It has been postulated that the increase in oxygen uptake is associated with phagocytosis
and is related to the subsequent process whereby the ingested microorganisms are killed. It has become apparent that biological systems are
able to convert oxygen into compound of great reactivity. The superoxide
anion (O_2^-), is capable of undergoing either oxygenation to O_2 or reduction to H_2O_2 with the liberation of large amounts of energy. Human peripheral blood polymorph nuclear leukocytes (PMN) generate superoxide anion (O_2^-) when exposed to apropriate stimuli. This highly reactive
molecule may be involved in bacterial killing. The nature of the O_2^--gen-
erating system and the mechanisms whereby its activity maybe enhanced,
however, are not well known. Evidence for the production of O_2^- by phagocytosing leukocytes have been put forward. *Babior* et al. [1] found this
superoxide radical as well at the site of inflammation as in the suspend-
ing medium. Superoxide generation in synovial fluid of injured joints has
been observed. The hydroxyl radical (OH·) a product of the reaction between superoxide and hydrogen peroxide is probably one of the most
toxic of the superoxides. Exact mechanism whereby leukocytes release
superoxides is not well known. The leucotriene LTB_4 is one of the compounds able to trigger superoxide release from leukocytes. Another compound with similar properties regarding superoxide generation is the anaphylatoxin C5a, which is inactivated by internalization into leukocytes

following desargination. In this process C5a is able to trigger superoxide release [2].

This report evaluates a possible relation between H_2O_2 release in the presence of increased C5a concentrations and its possible impact on the development of shock.

Material and Methods

Patients. 26 patients were studied who were in shock due to sepsis (n = 7), occlusion of aorta during surgery for ruptured aneurysms (n = 7), ischemia of intestine (n = 5) and hip surgery with fixation of the prothesis with methylmeta-acrylate (n = 7).

In vitro. Fresh serum from human volunteers was incubated with increasing amount of *E. coli* (n = 8).

Methods for Anaphylatoxin Determination. Blood was drawn and one portion was allowed to clot and the supernatant was removed and kept at –80 °C. Another part of the blood was immediately placed in ethylenediaminetetraacetic acid (EDTA) and centrifuged. The cells were removed and plasma was frozen at –80 °C until further processing. Plasma levels of C5a antigen from patients and in vitro studies were determined with a radioimmunoassay [3].

Preparation of Leukocyte Suspension. Leukocyte suspensions containing approximately 80 % PMN were prepared from heparinized venous blood (10 U/ml) obtained from healthy adult donors by employing standard techniques of dextran sedimentation and hypotonic lysis of erythrocytes [4]. Purified preparations of PMN were obtained by means of Ficoll-Hypaque gradients to avoid contamination of platelets and erythrocytes.

Determination of O_2. Duplicate reaction mixtures containing leukocytes (5×10^6 PMN) were used. The generation of O_2-dependent cytochrome c reduction was determined by the method of *Babior* et al. [1]. For studies of the effect of *E. coli*, 5×10^6 or 5×10^8 of *E. coli*, was added. The reaction mixture was accompanied by a control prepared with the same leukocytes suspension, identically in all respects except for the omission of the bacterias. Once prepared the reaction mixtures were stored on ice. Before incubation 1.5 ml of each reaction mixture was reserved at 0 °C for use as reference. The remainder was incubated for 15 min at 37 °C. The reactions were terminated by placing the vessels in melting ice. After centrifugation of blanks and incubating mixtures at 1,500 g for 5 min at 4 °C to remove leukocytes, cytochrome c and bacterias, the cytochrome c reduction was determined by measuring the absorbance of the incubated supernate at 550 nm using the unincubated blank as reference. E (ferrocytochrome c – ferricytochrome c) at 550 nm was taken as 15.5 mM.

Methods for in vitro Anaphylatoxin Experiments. Fresh serum alone or with *E. coli* in increasing amounts were incubated at body temperature for 15 min. The experiments were run in duplicates. The difference between C5a in serum incubated alone and with stimulating agents present were calculated.

Analysis of Data. Mean values and standard error were determined by standard methods. Pitmans test was used for comparison of groups. A p-value of < 0.05 was used for asigning statistical significance.

Results

In patients with shock due to sepsis, aorta occlusion, mesenterial ischemia or in association with hip replacement where methylmeta-acrylate has been used for fixation of the prothesis C5a levels were 5.2, 3.1, 4.8 and 4.0 times elevated compared to the initial normal plasma concentration.

Following incubation at body temperature for 15 min of serum alone or with 5×10^6 of 5×10^8 *E. coli*, C5a was 2.4 and 3.1 times increased compared to in serum alone. Corresponding figures for superoxide-dependent cytochrome *c* reduction per 5×10^6 PMN per 15 min were 3.1 and 5.2 nmol increased compared to serum incubated without bacterias.

Discussion

Increased intracellular oxygen radical production has been postulated as one contributing factor to the inflammatory response. When inflammation is extensive superoxide generation by bacteria, ischemic tissue or methylmeta-acrylate might be one of the grave threats. The correlation in time between C5a and superoxide generation indicate a possible relation. As cortisone can block complement activation this could be one possible way to reduce the effect of C5a generation and possibly also superoxide release [5, 6]. The effect of corticosteroids in septic shock and during rejection of a transplant could partly be explained in this way.

References

1   Babior, B. M.; Kipnes, R. S.; Curnette, J. T.: Biological defense mechanisms. The production by leukocytes of superoxide, a potential bactericidal agent. J. clin. Invest. *52:* 741–744 (1973).
2   Goldstein, I. M.; Brai, M.; Osler, A. G.; et al.: Lysosomal enzyme release from human leukocytes: mediation by the alternative pathway of complement activation. J. Immun. *111:* 33–37 (1973).
3   Wagner, J. L.; Hugli, T. E.: Radioimmunoassay for anaphylatoxins: a sensitive method for determining complement activation products in biological fluids. Analyt. Bichem. *136:* 75–88 (1984).
4   Boyum, A.: Isolation of mononuclear cells and granulocytes from human peripheral blood. Scand. J. clin. Invest. *21:* suppl. 97, pp. 77–89 (1968).

5   Packard, B. D.; Weiler, J. M.: Steroids inhibit activation of the alternative amplifi-
     cation pathway of complement. Infect. Immunity *40:* 1011–1014 (1983).
6   Fuenfer, M. M.; Carr, E. A.; Polk, H. C.: The effect of hydrocortisone of superox-
     ide production by leukocytes. J. surg. Res. *27:* 29–35 (1979).

M. Heideman, MD, PhD, University of Göteborg, Department of Surgery I,
Sahlgren's Hospital, S-413 45 Göteborg (Sweden)

Novelli, Ursini (eds.), Oxygen Free Radicals in Shock. Int. Workshop,
Florence 1985, pp. 60–65 (Karger, Basel 1986)

Lipid Peroxidation-Induced Mitochondrial Damage: Analogies with Ischaemic and Reperfusion Injury

*C. Ceconi[a], S. Curello[a], G. M. Boffa[a, 1], A. Cargnoni[a], A. Albertini[b],
R. Ferrari[a]*

[a]Cattedra di Cardiologia e [b]Cattedra di Chimica, Università degli Studi di
Brescia, Italy

Recent evidence suggests that, although there are multiple components to myocardial ischaemic and reperfusion damage, oxygen-derived free radicals may play a role in the pathophysiology of myocardial injury induced by ischaemia and/or post-ischaemic reperfusion [2–4]. The basic premise for the involvement of oxygen in reperfusion damage is that ischaemia has altered the defence mechanisms against oxygen toxicity. Our previous work [4] has demonstrated that severe ischaemia results in a marked and specific decline of mitochondrial (Mn-dependent) superoxide dismutase activity.

It is well known that heart mitochondria are able to produce reactive species of oxygen such as superoxide radical, hydrogen peroxide and hydroxyl radicals [5, 6]. It is conceivable, therefore, that the readmission of oxygen with reperfusion to mitochondria lacking in their natural defences against oxygen toxicity could result in an oxidative damage leading to structural and functional alterations.

In particular the unsaturated fatty acids of mitochondrial membranes as well as membrane-bound proteins containing oxidable amino acids are susceptible to free radical attack and the resultant damage may increase membrane permeability. This, in turn, would cause severe alterations of mitochondrial metabolic processes. In the present study, we examined the vulnerability to free radical injury of isolated rabbit heart mi-

---

[1] Dr. *Gianni Boffa,* Ricercatore of the Cattedra di Cardiologia, University of Padova, is on study leave for a year at the University of Brescia and he is supported by a scholarship from Fondazione V. Tonolli (Verbania).

tochondria incubated in the presence of ferrous ions as catalists of lipid peroxidation.

As endpoints of damage we have monitored the amount of lipid peroxidation products (malon dialdehyde) and the changes in mitochondrial function (measured in terms of oxygen utilization and Ca^{2+} transport).

Materials and Methods

Adult male rabbits were used. The hearts were excised and the mitochondria isolated by the method of *Sordahl* et al. [7] except that ethylenediaminetetra-acetic acid (EDTA) was present only in the homogenization media. The incubations with ferrous ions were carried out at 20 °C for 30 min in 105 mM KCl, 10 mM Tris, 10 mM Hepes buffer, pH 7.2. Ferrous ions, as $FeSO_4$, were dissolved in a small amount of deoxygenated 0.1 N HCl and added, within 4 min, to the mitochondrial suspension (1 mg/ml) to the final concentration of 0.2 mM.

Malondialdehyde (MDA) concentration was measured as thiobarbituric acid reactive material, in presence of 0.01 % butylated hydroxytoluene as described by *Buege and Aust* [8]. Mitochondrial oxygen consumption was determined by means of a Clark type electrode and Ca^{2+} transport by means of a selective Ca^{2+} electrode [9]. Protein determination was carried out according to the method of *Bradford* [10] using bovine serum albumin as standard. Data are expressed as mean + SE of n experiments. Statistical analysis was performed by Student's t-test, taking p = 0.05 as the limit of significance.

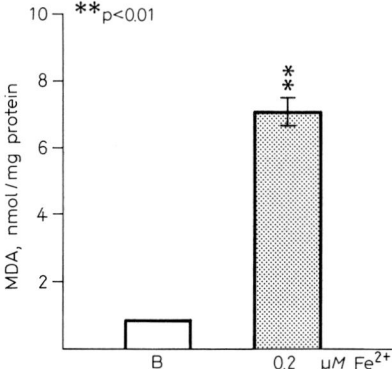

Fig. 1. Ferrous ions-induced formation of malondialdehyde: each value is the mean + SE of six separate experiments.

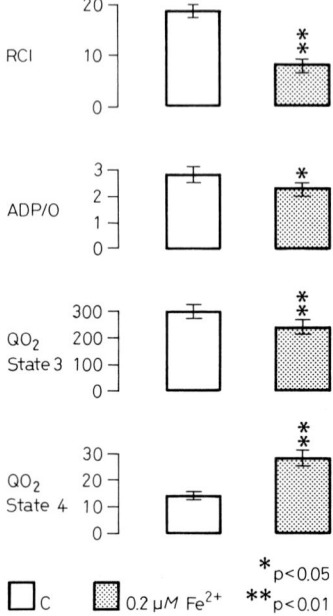

Fig. 2. Mitochondrial oxidative phosphorylation was monitored at the constant temperature of 25 °C using a Clark type oxygen electrode. Incubation medium consisted of 250 mM sucrose, 3 mM glutamate, 3mM KH$_2$PO$_4$; O.4 mM EDTA, 0.25 mM ADP, pH 7.4. 1.25 mg of mitochondrial protein were used in a final volume of 2.0 ml. The following indices have been monitored: Respiratory control index (RCI): the ratio between nmol of O$_2$ utilized in 1 min in presence and in absence of ATP. QO$_2$: n atoms of oxygen utilized in 1 min in the presence or in the absence (QO$_2$/4) of ADP. ADP/O: the ratio between nmol of ADP phosphorylated and the n atoms of oxygen utilized. Each value is the mean + SE of six separate experiments.

Results

Figure 1 shows the extent of lipid peroxidation induced by Fe^{2+} ions in our experimental conditions. Clearly 30 min of incubation with Fe^{2+} resulted in a massive accumulation of MDA.

Figure 2 shows the effect of lipid peroxidation on the function of the same mitochondria respiring with glutamate as substrate. Lipid peroxidation induced a decline of all parameters of mitochondrial oxygen consumption, although the respiratory control index (RCI) and the oxygen consumption in the presence of adenosine triphosphate (ATP; QO$_2$/3) were particularly depressed.

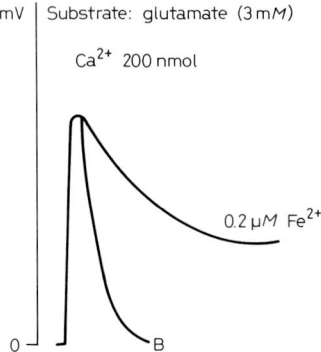

Fig. 3. Mitochondrial calcium transport was monitored using a selective Ca^{2+} electrode. The incubation medium contained 105 mM KCl, 10 mM Tris, 10 mM Hepes, 3 mM KH_2PO_4, 3 mM glutamate, pH 7.2. In a final volume of 1.5 ml, calcium transport of mitochondria (1 mg of protein) was started adding to the medium 200 nmol of $CaCl_2$. All experiments were carried out 6 times with a high degree of reproducibility. Only one representative experiment is shown.

Figure 3 reports a typical tracing on calcium movement obtained by a selective calcium electrode. A decrease of Ca^{2+} concentration in the medium means a transport of Ca^{2+} into the mitochondria. When the control, non-iron-incubated mitochondria were used, the Ca^{2+} content of the reaction medium returned approximately to the level seen before the Ca^{2+} addition, whilst in the iron-incubated mitochondria the rate of transport and capacity of retaining calcium were severely reduced.

Discussion

The role of ferrous ions in the propagation and chain branching of lipid peroxidation is well documented [11, 12]. The choice of this incubation system, which is less efficient in respect to those with EDTA or adenosine 5'-diphosphate (ADP) chelated iron, or with reducing substances, has enabled us to perform on the same preparation measurements of mitochondrial oxygen consumption and of calcium transport.

Our data demonstrate that lipid peroxidation results in a marked compromise of mitochondrial function, inducing two main phenomena: uncoupling of oxidative phosphorylation and inhibition of respiration. Likely, the uncoupling effect is explained by an increase in proton permeability of the membrane lipid bilayer and it is well documented by the severe reduction of RCI and ADP/O. This also explains the reduced calcium accumulation capacities of the same mitochondria, as transport is driven by electron. The effect of ferrous ion-induced lipid peroxidation on the inhibition of mitochondrial respiration is more difficult to explain, but it is well documented by the reduction of oxygen consumption during state 3 ($QO_2/3$).

It is worthwhile to note that similar alterations of mitochondrial function and of mitochondrial Ca^{2+} transport have been previously reported by other authors employing different systems of peroxidative damage [13, 14]. In addition, we would like to underline that the described alterations in mitochondrial function are very similar to those observed in mitochondria isolated after post-ischaemic reperfusion [15]. This might confirm a role for free radicals in the determination of ischaemic and reperfusion damage.

Acknowledgements

This study was supported by CNR Grants No. 82.02337.56 and No. 82.02739.04. We are very grateful for the assistance of Miss *Ornella del Ciello* in the typing of the manuscript.

References

1   Poole-Wilson, P. A.; Harding, D. P.; Bourdillon, P. D. V.; Tones, M. A.: Calcium out of control. J. mol. cell. Cardiol. *16:* 175–187 (1984).

2   Meerson, F. Z.; Kagan, V. E.; Kozov, Y. P.; Belkina, L. M.; Arkhipenko, Y. V.: The role of lipid peroxidation in pathogenesis of ischaemic damage and the antioxidant protection of the heart. Basic Res. Cardiol. *77:* 465–485 (1982).

3   Rao, P. S.; Cohen, M. S.; Mueller, H. S.: Production of free radicals and lipid peroxides in early experimental ischaemia. J. mol. cell. Cardiol. *15:* 713–716 (1983).

4   Ferrari, R.; Ceconi, C.; Curello, S.; Guarnieri, C.; Caldarera, C. M.; Albertini, A.; Visioli, O.: Oxygen-mediated myocardial damage during ischaemia and reperfusion: role of the cellular defences against oxygen toxicity. J. mol. cell. Cardiol. *17:* 937–945 (1985).

5   Nohl, H.; Hegner, D.: Do mitochondria produce oxygen radicals in vivo? Eur. J. Biochem. *82:* 563–567 (1978).

6   Nohl, H.: The biochemical mechanism of the formation of reactive oxygen species in heart mitochondria; in Caldarera, Harris, Advances in studies on heart metabolism, pp. 413–421 (Clueb, Bologna 1982).

7   Sordahl, L.; McCollum, W. B.; Wood, W. G.; Schwartz, A.: Mitochondrial and sarcoplasmic reticulum function in cardiac hypertrophy and failure. Am. J. Physiol. *224:* 497–502 (1973).

8   Buege, J. A.; Aust, S. A: Microsomal lipid peroxidation; in Fleisher, Packer, Methods in enzymology, No. 52, pp. 302–310 (Academic Press, New York 1978).

9   Brand, M. D.; Chen, C. H.; Lehninger, A. L.: Stoichiometry of H^+ ejection during respiration-dependent accumulation of Ca^{2+} by rat liver mitochondria. J. biol. Chem. *251:* 968 (1980).

10 Bradford, M. M.: A rapid and sensitive method for the quantitation of microgram quantities of protein utilizing the principle of protein-dye binding. Analyt. Biochem. *72:* 246–254 (1976).

11 Vladimirov, Y. A.; Olenev, V. I.; Syslova, T. B.; Cheremisina, Z. P.: Lipid peroxidation in mitochondrial membrane. Adv. Lipid Res. *17:* 173–249 (1980).

12 Halliwel, B.; Gutteridge, J. M.: Oxygen toxicity, oxygen radicals, transition metals and disease. Biochem. J. *219:* 1–14 (1984).

13 Guarnieri, C.; Ceconi, C.; Muscari, C.; Flamigni, F.: Influence of oxygen radicals on heart metabolism; in Caldarera, Harris, Advances in studies on heart metabolism, pp. 423–431 (Clueb, Bologna 1982).

14 Harris, E. J.; Both, R.; Cooper, M. B.: The effects of superoxide generation on the ability of mitochondria to take up and retain calcium. FEBS Lett. *146:* 267–272 (1982).

15 Ferrari, R.; Williams, A.; Di Lisa, F.: The role of mitochondrial function in the ischaemic and reperfused myocardium; in Caldarera, Harris, Advances in studies on heart metabolism, pp. 245–255 (Clueb, Bologna 1982).

Dr. C. Ceconi, Cattedra di Cardiologia, Università degli Studi di Brescia,
c/o Spedali Civili, I-25100 Brescia (Italy)

Novelli, Ursini (eds.), Oxygen Free Radicals in Shock. Int. Workshop, Florence 1985, pp. 66–73 (Karger, Basel 1986)

Aspects of Mitochondrial Electron Transport Chain Oxidative Damage and Oxygen Free Radical Production during the Course of Circulatory Shock in Humans[1]

A. Gasparetto, G. G. Corbucci, M. Antonelli, M. Bufi, R. A. De Blasi, G. Crimi

Institute of Anaesthesiology and Resuscitation, University La Sapienza, Policlinico Umberto I°, Rome, Italy

Introduction

It is well known that the cellular pathogenic development of circulatory shock is mainly connected to a very low oxygen tension, due to tissue hypoxia, and there is considerable evidence supporting a key role for oxygen free radicals in the mitochondrial oxidative damage [1, 10]. Our results on humans, previously reported [1], such as the data obtained in experimentally induced shock in animals [10], suggest a linear relationship between the marked decrease of electron transport chain (ETC) oxidative capacity and superoxide dismutase content, hydroperoxide levels, and oxidized/reduced glutathione investigated during the course of syndrome. As regards the oxygen free radical generation, the production of superoxide radicals at the NAD(P)H dehydrogenases as well as at ubiquinone-cytochrome c sites, enhanced by one-electron reduction of molecular oxygen, could occur in conditions characterized by a significant decrease on the ETC oxidative capacity [3]. In this pathogenic situation the importance of α-tocopherol in preventing the free radical attack on membrane phospholipids has been widely stressed [6] and similar proper-

[1] This work conforms to ethical considerations for research on hospital patients and was approved by the Human Experimentation Review Committee of the Anaesthesia and Resuscitation Departments, University of Rome.

ties were first ascribed to ubiquinone by *Mellors and Tappel* [9]. More recently the attention was focused on the antioxidant role which ubiquinol can play in mitochondria peroxidative disorders and the ubiquinol (UQH_2) activity could be ascribed both to a reaction with lipid free radicals and to a structural effect on the lipid bylayer [5]. In order to evaluate the suggested relationship between the mitochondrial ETC oxidative damage and the role of oxygen free radical production, during the course of circulatory shock in humans, in the present study we investigated cytochrome *c* oxidase (COX), succinate cytochrome *c* reductase (SCR) activities and coenzyme Q content in respect to α-tocopherol levels, assumed as antioxidant agent and then as indirect index of peroxidative processes.

Materials and Methods

Clinical Study. 10 shock patients, admitted to the Intensive Care Unit of the Policlinico Umberto I^0 in Rome, were studied. Cardiogenic shock was diagnosed by clinical general criteria and the haemodynamical and biochemical parameters were examined at 3-hour intervals expressed in tables I–IV as T_1, T_2, T_3, T_4, T_5, subsequent to admission to ICU (T_0). During the course of shock the patients were monitored by Swan-Ganz catheter 7 Fr, connected to Bentley Trantec mod. 800 transducers and to the pressure system Kontron Roche 244 (Kontron Roche, Basel, CH). The cardiac index was determined in triplicate by the thermodilution method by an Edwards computer 9250 (Edwards Lab., London, UK).

Biochemical Study. The biochemical parameters were investigated on skeletal muscle samples (80 mg) taken by needle biopsies from the vastus lateralis muscle, as described by *Edwards* et al. [2], as soon as possible after the first diagnosis of cardiogenic shock and at the time intervals described in the clinical study. The tissue samples were analysed within 15 min after collection. Control values were obtained by taking muscle biopsies from 10 healthy volunteers.

Biochemical Methods. ETC oxidative capacity was assayed by COX and SCR activities using the method proposed by *Gohil* et al. [4]. The spectrophotometric analyses were carried out by a Shimadzu UV 3000 spectrophotometer. Coenzyme Q homologs and α-tocopherol levels were determined as described by *Takada* et al. [11] and following the modifications proposed by *Antonelli* et al. [this book].

Data obtained from the activities found to be present in the first biopsy taken after the onset of shock (T_0) were compared to those found in healthy subjects by paired t-test and the p-value determined. Progressive changes in activities present in the patients during the course of shock were evaluated by t-test analysis. The value shown for each parameter measured is the average of three separate determinations. The coefficient of variation of each series of measurements is given in the tables, where statistical estimations are also reported.

Table I. Haemodynamic parameters for 10 circulatory shock patients

Parameter (abbreviation)	Normal values	Shock patients, mean ± SEM					
		T_0	T_1	T_2	T_3	T_4	T_5
Heart rate (HR), bpm^{-1}	60–80	95.09 ±4.13	93.63 ±3.62	93.36 ±3.31	99.15 ±0.16	102.6 ±6.01	105 ±6.04
Mean artery pressure (MAP), mm Hg	85–95	64.18 ±6.71	66.36 ±6.42	68.09 ±6.30	74 ±7.24	82.6 ±7.63	74.77 ±5.84
Mean pulmonary artery pressure (MPAP), mm Hg	11–17	19.90 ±2.68	21.09 ±2.99	23.63 ±2.23	26.63 ±2.4	28.8 ±2.17	26.44 ±2.00
Cardiac index (CI), $l \cdot min^{-1} \cdot m^{-2}$	3.2–3.8	2.81 ±0.29	2.89 ±0.31	3.16 ±0.32	3.72 ±0.39	3.43 ±0.38	3.73 ±0.30
Total pulmonary resistance index (TpRI), $dyne \cdot s \cdot cm^{-5} \cdot m^{-2}$	250–400	637.5 ±135.40	661.37 ±127.8	656 ±73.68	768.2 ±74.88	766.87 ±86	372.87 ±60.27
Total peripheral resistance index (TPRI), $dyne \cdot s \cdot cm^{-5} \cdot m^{-2}$	1,800–2,200	1,963.6 ±192.26	2,071.5 ±193.4	2,005.3 ±285.4	2,162.5 ±272.7	2,084.1 ±290.7	1,742.87 ±215.17

Table II. Cytochrome *c* oxidase (COX) and succinate cytochrome *c* reductase (SCR) activities in shock patients

T_0	T_1	T_2	T_3	T_4	T_5
COX, μmol cytochrome *c* oxidized·min⁻¹·g tissue⁻¹[a] (mean ± SEM)					
5.31±0.93	2.37±1.4	1.46±0.79	1.54±0.93	1.36±0.79	0.71±0.24
SCR, μmol cytochrome *c* reduced·min⁻¹·g tissue⁻¹[b] (mean ± SEM)					
0.44±0.15	0.17±0.07	0.18±0.10	0.18±0.12	0.16±0.11	0.06±0.02

[a] Normal values: 16.09±0.30 (n = 10). When compared to the value of the first biopsy taken (T_0), subsequent values showed no significant differences by a paired Student's t-test. The coefficient of variation was never greater than 10%.
[b] Normal values: 3.33±1.33 (n = 10). Compared to the value of the first biopsy taken (T_0) by paired Student's t-test, those values obtained from subsequent biopsies showed no significant differences during the course of shock. The coefficient of variation of these measurements was 1.8%.

Results

The clinical and haemodynamical parameters and derived data from 10 cardiogenic patients are summarized in table I. The data confirm the initial diagnosis of circulatory shock, suggesting a positive answer to the administered therapy, based on the restoration of the blood buffer with $NaCO_3^-$, reestablishment of the circulating volume as necessary, vasoactive substances according to haemodynamic considerations and ventilation as required. After the onset of shock COX activity is significantly lower in shock patients than that present in normal subjects (table II, T_0) and a similar decrease is observed when the capacity of SCR is determined (table II, T_0). After the initial fall in oxidative capacity, COX and SCR activities show a further decrease after a 3-hour interval (table II, T_1). During the course of syndrome the ETC oxidative capacity remains relatively constant, showing no further significant fall as condition progressed. In relative contrast with this pattern, the total coenzyme Q_{10}, after a considerable decrease after the onset of shock (table III, T_0), shows a trend comparable with that observed for COX and SCR activities. Simultaneous determinations of oxidized and reduced coenzyme Q homologues were carried out to determine whether or not the fall in COX activity and in the total Q_{10} content affect the $Q_{10}/Q_{10}H_2$ ratio.

Table III. Coenzyme Q contents in shock patients

T_0	T_1	T_2	T_3	T_4	T_5
Coenzyme Q_{10}, $\mu g \cdot g$ tissue^{-1} [a] (mean ± SEM)					
15.29±1.70	13.72±2.11	15.42±2.14	14.31±2.40	14.80±2.66	11.13±3.87
Coenzyme $Q_{10}H_2$, $\mu g \cdot g$ tissue^{-1} [b] (mean ± SEM)					
11.19±3.29	16.79±2.83	15.75±3.32	14.41±3.63	15.96±3.48	17.95±6.24
Total coenzyme Q_{10}, $\mu g \cdot g$ tissue^{-1} [c] (mean ± SEM)					
26.48±2.07	30.50±1.61	31.16±2.37	28.77±3.11	30.71±2.77	28.95±2.55
Coenzyme $Q_{10}/Q_{10}H_2$, ratio [d] (mean ± SEM)					
4.99±2.07	3.43±1.80	2.09±1.10	2.54±1.08	1.16±0.34	0.32±0.05
Coenzyme Q_9, $\mu g \cdot g$ tissue^{-1} [e] (mean ± SEM)					
0.454±0.06	0.323±0.033	0.409±0.045	0.463±0.084	0.442±0.084	0.277±0.033

[a] Normal values: 34.65±3.20 (n=10). Paired t-test analysis of the data gave the following statistical evaluation: T_0vs normal: <0.001; T_1vs normal: <0.001; T_2vs normal: <0.001; T_3vs normal: <0.001; T_4vs normal: <0.001; T_5vs normal: <0.01. The coefficient of variation was 8% or less.

[b] Normal values: 6.48±0.79 (n=10). Paired t-test analysis of the data gave the following statistical evaluation: T_0vs normal: ns; T_1vs normal: <0.001; T_2vs normal: <0.01; T_3vs normal: <0.02; T_4vs normal: <0.01; T_5vs normal: <0.01. The coefficient of variation was 10%.

[c] Normal values: 41.38±3.18 (n=10). Paired t-test analysis of the data gave the following statistical evaluation: T_0vs normal: <0.01; T_1vs normal: <0.02; T_2vs normal: <0.05; T_3vs normal: <0.02; T_4vs normal: <0.05; T_5vs normal: n.s. The coefficient of variation was 10% or less.

[d] Normal values: 7.15±1.75 (n=10).

[e] Normal values: 0.99±0.194 (n=10). Paired t-test analysis of the data gave the following statistical evaluation: T_0vs normal: <0.01; T_1vs normal: <0.01; T_2vs normal: <0.01; T_3vs normal: <0.02; T_4vs normal: <0.02; T_5vs normal: <0.05. The coefficient of variation was 6%.

When compared to that present in normal subjects, the cellular hypoxia due to circulatory shock causes a marked increase in the reduced form ($Q_{10}H_2$) and a reciprocal fall in the oxidized form (Q_{10}) as represented in table III. This pattern is particularly significative at time T_1 and the same trend can be observed for the coenzyme Q_9 levels (table III). α-Tocopherol is thought to function as a free radical scavenger and the effect of

Table IV. α-Tocopherol levels during the course of circulatory shock

T_0	T_1	T_2	T_3	T_4	T_5
α-Tocopherol, mg · g tissue^{-1} [a] (mean±SEM)					
31.74±6.04	21.36±3.29	26.75±4.65	31.09±9.89	29.60±6.14	16.22±4.68

[a] Normal values: 15.09±1.35 (n=10). Paired t-test analysis of the data gave the following statistical evaluation: T_0vs normal: 0.01; T_1vs normal: n.s.; T_2vs normal: 0.01; T_3 vs normal: 0.05; T_4vs normal: 0.01; T_5vs normal: n.s. The coefficient of variation was 10%.

the circulatory shock was examined to relate the α-tocopherol changes to those found in coenzyme Q_{10} levels and in the ETC enzymic activities. The cellular content of α-tocopherol shows a significant increase after the onset of shock (table IV, T_0), when compared to that present in normal muscle tissue, while the fall in the mitochondrial oxidative activity seems to involve the α-tocopherol levels during the course of syndrome (table IV).

Discussion

In agreement with our previous report [1], the results of this study confirm that the mitochondrial oxidative capacity shows a considerable fall during the circulatory shock and this pattern appears to be independent from the interpatient variations. When the results are expressed with respect to tissue wet weight, the significant loss of tissue oxidative activity appears to play a key role in the pathogenic development of septic (last phase) and cardiogenic shock [unpublished data]. Since the activity of COX does not require the presence of cofactors, the inactivation seen can be due to damage to some or all of its components (subunits), even if the high affinity of COX for oxygen can guarantee the survival rate of the enzyme. It is well known that the univalent and bivalent reduction of molecular oxygen give rise to superoxide and hydrogen peroxide, respectively [8], and the inactivation of COX is suggested as being involved in superoxide and hydrogen peroxide production by an autoxidable ubisemiquinone generation [3]. In light of these findings, it appears relevant to note that after the onset of shock (T_1) a relationship exists between the marked loss of COX activity and the increase of $Q_{10}H_2$ content, which might act as an antioxidant. This fact appears more evident when $Q_{10}/Q_{10}H_2$ ratio is analyzed. Several possibilities exist to explain these changes observed during the course of circulatory shock. It may be that the COX and coenzyme Q_{10} lowered activities result from a primitive metabolic adaptive changes due to the cellular oxygen tension. Alternatively, it may be that the structural and functional damage of mitochondrial inner membrane phospholipids inactivate these enzymic complexes. This does not rule out the possibility that the above-mentioned mechanisms are in some way connected, by an initial mono- or bivalent reduction of molecular oxygen, with the subsequent superoxide and hydrogen peroxide generation which then characterizes the inner membrane lipoperoxidative damage. Our previous results showed that the muscle cells have a reduced capacity to remove superoxide, due to a fall in superoxide dismutase content observed during the course of shock [1]. As regards the endogenous oxygen free radical scavengers, the ability of superoxide dismutase in preventing the peroxidative chain reactions is well know [7], and there is a good body of evidence that α-tocopherol can play a crucial role in maintaining the physical properties of phospholipid bylayer membranes [12]. In the present study the simultaneous determination of α-tocopherol and $Q_{10}H_2$ levels shows an opposite trend. After the onset of shock the decrease of α-tocopherol may indicate a consumption of this antioxidant by scavenging free radicals in biomembranes, while $Q_{10}H_2$ content increases. This could mean that $Q_{10}H_2$ may not be involved in scavenging free radicals in the initial step of lipid peroxidation, while α-tocopherol, in its chain-breaking function of free radical reactions, remains active during the course of shock. Finally, it is important to point out that after the onset of shock this scavenger shows raised levels, when compared to normal values, and this fact may be due to higher α-tocopherol turnover, as a response to cellular oxygen free radical increased amounts. From the results presented here, we suggest that, during the course of circulatory shock in humans, the oxidative damage of some components of mitochondrial ETC can play a crucial role in the pathogenesis of syndrome. Although more definitive proofs of oxygen free radical mitochondrial source are required, the presented data provide a suggestive evidence that the mitochondrial inner membrane peroxidative damage, activated by the oxygen free radical intermediates, may characterize the reversible and/or irreversible phases of circulatory shock.

References

1   Corbucci, G. G.; Gasparetto, A.; Candiani, A.; Crimi, G.; Antonelli, M.; Bufi, M.; De Blasi, R. A.; Cooper, M. B.; Gohil, K.: Shock-induced damage to mitochondrial function and some celluar antioxidant mechanisms in humans. Circul. Shock 15: 15–26 (1985).

2   Edwards, R. H. T.; Young A.; Wiles, M.: Needle biopsy of the skeletal muscle in the diagnosis of myopathy and in the clinical study of muscle function and repair. New Engl. J. Med. 302: 261–271 (1980).

3   Forman, H. J.; Boveris, A.: Superoxide radical and hydrogen peroxide in mitochondria; in Boveris, Free radicals in biology, vol. V, pp. 65–87 (Academic Press, Orlando 1982).

4   Gohil, K.; Jones, D. A.; Edwards, R. H. T.: Analysis of mitochondrial function with techniques applicable to needle muscle biopsy samples. Clin. Sci, 66: 121–128 (1981).

5   Landi, L.; Cabrini, L.; Sechi, A. M.; Pasquali, P.: Antioxidative effect of ubiquinones in mitochondrial membranes. Biochem. J. 222: 463–466 (1984).

6   Marubayashi, S.; Dohi, K.; Yamada, K.; Kawasaki, T.: Changes in the levels of endogenous coenzyme Q homologs, α-tocopherol, and glutathione in rat liver after hepatic ischemia an reperfusion, and the effect of pretreatment with coenzyme Q_{10}. Biochim. biophys. Acta 797: 1–9 (1984).

7   McCord, J. M.; Keele, B. B.; Fridovich, I.: An enzyme-based theory of obligate anaerobiosis. The physiological function of superoxide dismutase. Proc. natn. Acad. Sci. USA 68: 1024–1027 (1971).

8   McCord, J. M.: The superoxide free radical: its biochemistry and pathophysiology. Surgery, St Louis 3: 412–414 (1983).

9   Mellors, A.; Tappel, A. L.: Antioxidative effect of ubiquinones. J. biol. Chem. 240: 3439–3446 (1966).

10   Parks, D. A.; Bulkely, G. B.; Granger, D. N.: Role of oxygen free radical in shock, ischemia and organ preservation. Surgery, St. Louis 94: 428–432 (1983).

11   Takada, M.; Ikenoya, S.; Yuzuriha, T.; Katayama, K.: Simultaneous determination of reduced and oxidized ubiquinones; in Packer, Oxygen radicals in biological system; methods in enzymology, No. 105, pp. 147–155 (Academic Press, Orlando 1984).

12   Takenaka, M.; Tatsukawa Y.; Dohi K.; Ezaki, H.; Matsukawa, K.; Kawasaki, T.: Protective effects of α-tocopherol and coenzyme Q_{10} on warm ischemic damages of the rat kidney. Transplantation 2: 137–141 (1981).

Prof. A. Gasparetto, Institute of Anaesthesia and Resuscitation, University La Sapienza, Policlinico Umberto I°, Viale del Policlinico, I-00161 Rome (Italy)

Novelli, Ursini (eds.), Oxygen Free Radicals in Shock. Int. Workshop,
Florence 1985, pp. 74–78 (Karger, Basel 1986)

The Role of Polymorphonuclear Leukocytes in Postischemic Delayed Hypoperfusion

Bjarne Grögaard, Ludwig Schürer, Bengt Gerdin, Karl-E. Arfors

Department of Experimental Medicine, Pharmacia AB, and the Department of
Surgery, University Hospital, Uppsala, Sweden

Clinical and experimental data on reversible cerebral ischemia have
shown that a period of circulatory arrest is followed by a brief period of
hyperperfusion and subsequently by progressive delayed hypoperfusion
(DHP) with a cerebral blood flow that is 30–50% of the normal in spite
of normalization of arterial blood pressure [*Siesjö*, 1978]. It is argued that
DHP might contribute to the final neurological outcome by leading to a
secondary ischemic impact to the brain. The aim of the present study was
to evaluate to what extent polymorphonuclear leukocytes (PMNL) contribute to DHP.

Materials and Methods

Fasted male rats of the Wistar strain, 300–400 g, were anesthetized, intubated and
maintained on N_2O/O_2 anesthesia under muscle relaxation and controlled ventilation.
Blood glucose concentrations, hematocrit and acid-base status were followed during
the experiment. Catheters were inserted into the tail artery and vein for blood pressure
recordings, blood sampling and infusions, into the right atrium for exsanguination, and
into the right brachial artery for blood sampling during ^{14}C-iodoantipyrine infusion.
Both carotid arteries were dissected free and prepared for clamping. EEG and body
temperature were continuously monitored.

Preparation of Antineutrophil Serum (ANS)
This was done as previously described by *Lundberg* et al. [1984]. The immuniza-
tion, however, took place in sheep. PMNL (50×10^6, 0.5 ml) were mixed with an equal
volume of Freund's complete adjuvant and injected i.m. every second week. Blood was
collected, whereafter the antiserum was processed as described by *Lundberg* et al.
[1984]. Normal sheep serum (NSS) for control experiments was collected prior to im-
munization and treated in the same way as the ANS.

Experimental Protocol

Rats were injected i.p. with 1 ml of ANS or NSS at −24 and −12 h. Orbital blood samples were taken for PMNL counts before treatment and before and after experimentation. Cerebral ischemia was produced by carotid artery clamping simultaneous with induction of hemorrhagic hypotension to 50 mm Hg and isoelectric EEG. Arfonad® was given on induction of ischemia to a total dose of 15 mg/kg body weight. After 15 min of ischemia, the clamps were released and the shed blood was retransfused. 60 min thereafter local cerebral blood flow (1CBF) was measured with the ^{14}C-iodoantipyrine technique [*Sakurada* et al., 1978; *Abdul-Rahman* et al., 1979]. 40 µCi of ^{14}C-iodoantipyrine in 1 ml of saline was infused i.v. for 45 s and blood samples were obtained every 5 s during ^{14}C-iodoantipyrine infusion. The animals were then killed by guillotine decapitation and the brain was immediately frozen in isopentane chilled to −50 to −60 °C.

Results

During ischemia the cortical structures and the caudoputamen and hippocampus exhibited near complete cessation of blood flow, while more caudally located regions showed more variable flows. 5 min after restitution of flow there was hyperperfusion, with 1CBF of more than 200% of the normal. 60 min thereafter the cortical flow was 30–50% of the normal, i.e. the ischemic insult in this model was followed by DHP.

Animals given ANS showed a pronounced decrease in the number of circulating PMNLs immediately before anesthesia (fig. 1). The PMNL count remained very low during the experimental procedure, in contrast

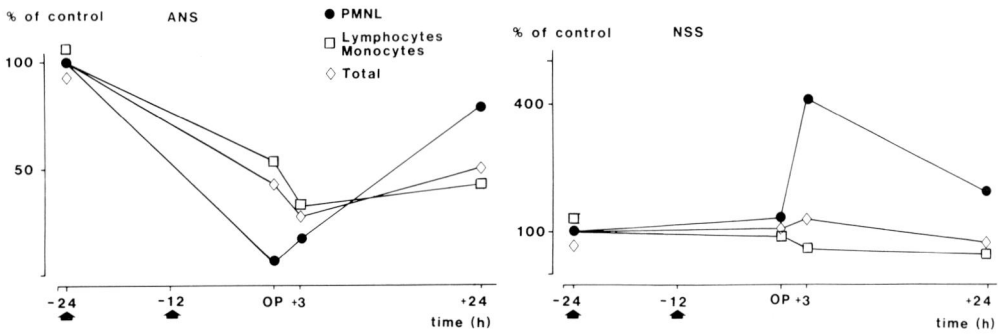

Fig. 1. Blood white cell counts in animals given ANS or NSS expressed in percent of the value at −24 h. Arrows indicate i.p. infusion of ANS or NSS.

ICBF ml/min g

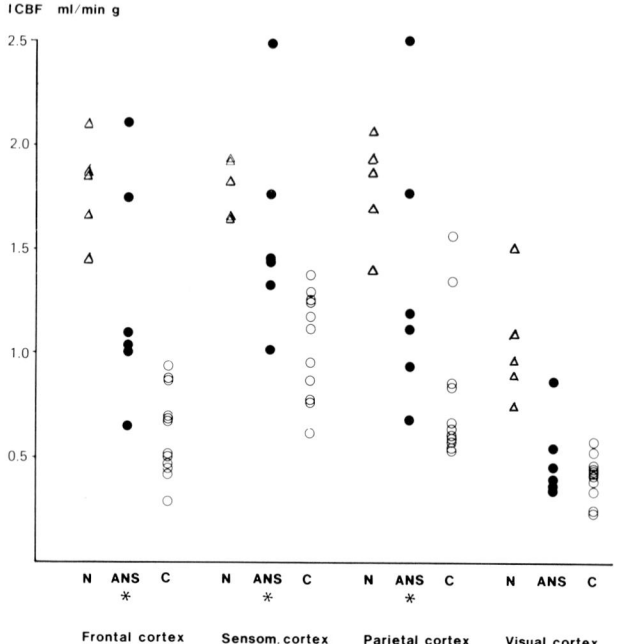

ICBF ml/min g

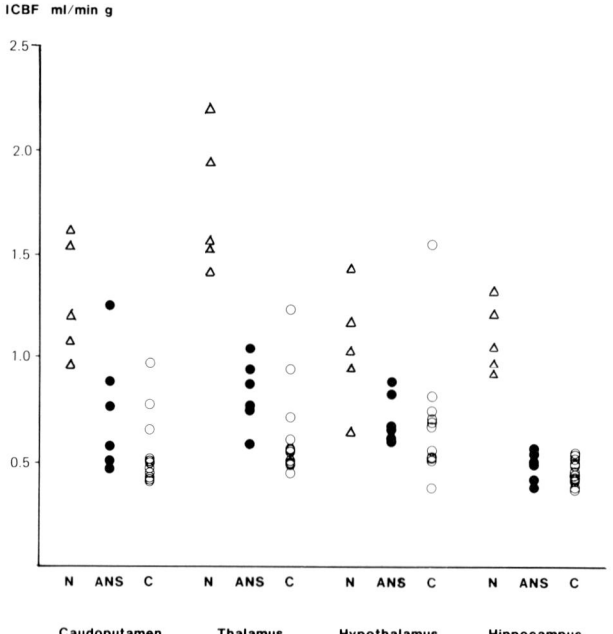

to that in animals given NSS, which was unaltered before experimentation and increased 4-fold during the experimental procedure.

In animals given ANS, lCBF was less disturbed during the postischemic DHP phase in the frontal, sensorimotor and parietal cortex than in animals not given ANS. In the visual cortex and subcortical structures there was no difference between the two experimental groups (fig. 2).

Discussion

The protective effect of ANS treatment on DHP in rostral cortical areas indicates that PMNLs in one way or another are involved in the flow disturbances. In the present study the animals were pretreated with ANS, which meant that both during and after ischemia the experimental groups differed. It cannot therefore be stated conclusively that the flow disturbance is due to an activity of PMNLs that is mediated after termination of the ischemic period. The existence of substantial hyperperfusion immediately after the ischemia, however, indicates that initially the vascular bed is not damaged.

The reason why the effect is only seen in rostral cortical structures but not in the visual cortex or in subcortical areas is not understood. It could either indicate that other mechanisms than PMNL plugging in the microcirculation are involved, or that the vascular bed in the cortical areas has a different architecture than that in the subcortical structures.

Previous studies on the mechanism of the delayed hypoperfusion have been focused on vascular smooth muscle supersensitivity to calcium, and therapeutic effects in various experimental models have been obtained by calcium antagonistic substances. It has also been proposed that swelling of perivascular astroglia causing extravascular compression may be of importance. The present results indicate, however, that PMNL activity is in some way involved. One possible mechanism is that an altered PMNL-endothelium interaction in the cerebral microcirculation causes mechanical obstruction of individual capillaries. This would be in accordance with findings in the myocardium [*Engler* et al., 1983] and in skeletal muscle [*Bagge* et al., 1980] presented by other groups.

Fig. 2. Local cerebral blood flow (lCBF) in various brain structures 60 min after termination of cerebral ischemia in animals pretreated and not pretreated with ANS (control group C). Group N is sham-operated animals. * $p < 0.05$ versus group C.

Electron microscopic studies have previously shown that delayed hypoperfusion is characterized by both astrocyte swelling and swelling of the individual endothelial cells. It is therefore possible that it is not intravascular plugging by PMNL as such that is the obstructing factor, but rather damage to the endothelium induced by activated PMNL.

Irrespective of the final mechanism involved, the results of the present study indicate that attempts to manipulate PMNL activation after ischemia could be a fruitful approach to combatting postischemic delayed hypoperfusion in the brain.

References

Abdul-Rahman, A.; Dahlgren, N.; Ingvar, M.; Rehncrona, S.; Siesjö, B. K.: Local versus regional cerebral blood flow in the rat at high (hypoxia) and low (phenobarbital anesthesia) flow rates. Acta physiol. scand. *106:* 53–60 (1979).

Bagge, U.; Amundson, B.; Lauritzen, C.: White blood cell deformability and plugging of skeletal muscle capillaries in hemorrhagic shock. Acta physiol. scand. *180:* 159–163 (1980).

Engler, R. I.; Schmid-Schönbein, G. W.; Pavelec, R. S.: Leukocyte capillary plugging in myocardial ischemia and reperfusion in the dog. Am. J. Path. *111:* 98–111 (1983).

Lundberg, C.; Lebel, L.; Gerdin, B.: Inflammatory reaction in an experimental model of open wounds in the rat. The role of polymorphonuclear leukocytes. Lab. Invest. *50:* 726–732 (1984).

Sakurada, O.; Kennedy, C.; Jehle, J.; Brown, J. D.; Carbin, G. L.; Sokoloff, L.: Measurement of local cerebral blood flow with iodo(^{14}C)antipyrine. Am. J. Physiol. *234:* H59–H66 (1978).

Siesjö, B. K.: Brain energy metabolism (Wiley & Sons, New York 1978).

B. Grögaard, MD, Department of Experimental Medicine, Pharmacia AB, S-751 82 Uppsala (Sweden)

Novelli, Ursini (eds.), Oxygen Free Radicals in Shock. Int. Workshop,
Florence 1985, pp. 79–82 (Karger, Basel 1986)

Inhibition of Oxidative Effects of Human Polymorphonuclear Granulocytes by Rosmarinic Acid

Kok P. M. van Kessel[a], Eric S. Kalter[b], Jan Verhoef[a]

[a] Laboratory of Microbiology, State University of Utrecht;
[b] University Hospital Sint Radboud, Department of Intensive Care, Nijmegen,
The Netherlands

Introduction

Ingestion of bacteria by human polymorphonuclear leukocytes (PMN) and perturbation of the cell membrane initiates a burst of oxygen consumption. This metabolic burst results in the formation of highly toxic oxygen species which play an important role in the PMN defense mechanisms [1]. As a consequence, these toxic oxygen products are released from the triggered PMN and tissue damage may occur at sites of infection and inflammation. The experimental drug rosmarinic acid (Nattermann, Cologne, FRG) was shown to possess anti-inflammatory properties in an animal model of the adult respiratory distress syndrome and multiple organ inflammation, based on complement-induced granulocyte activation [2]. Therefore, it was of interest to study in vitro the effects of rosmarinic acid on a number of human PMN functions, related to host defense mechanism and the production of oxygen radicals.

Methods and Results

We used human PMN, isolated from venous blood from healthy volunteers by dextran sedimentation and Ficoll-Paque density centrifugation. Cells were resuspended in Hanks' balanced salt solution after hypotonic shock of the residual erythrocytes as described [3]. All measurements were repeated at least 3 times.

Migration of PMN, directed towards chemotactic factors, was determined by the under agarose technique [4]. Agarose containing activated human pooled serum (C5a) served as a chemoattractant. After 18 h of incubation, no differences were observed in migration distance in the presence or absence of various concentrations of rosmarinic acid (final conc. from 1 to 500 μM) in the agarose.

The phagocytic capacity of PMN was tested by their ability to ingest [^{3}H] -thymidine radiolabeled *S. aureus* bacteria, a catalase positive strain, as described [3]. Bacteria were preopsonized by incubation in 5% human pooled serum as a source of both antibodies and complement, and added to PMN in a final bacteria to PMN ratio of 10:1. The percentage uptake after 12 min incubation at 37 °C was about 88% and was unaffected in the presence of various concentrations rosmarinic acid (1, 10, 100 and 500 μM). Also the subsequent intracellular killing of the ingested bacteria after 12 min phagocytosis was identical with or without the addition of rosmarinic acid and only 12% of the ingested bacteria appeared viable after osmotic lysis of the PMN as described [3].

An important parameter of the oxidative metabolism of the phagocytic cell is the generation of light emission or chemiluminescence. Reactive oxygen products are involved in the process of chemiluminescence by stimulated PMN as a consequence of membrane perturbation [1]. We studied the effects of rosmarinic acid on the luminol-dependent chemiluminescence response of PMN stimulated with opsonized *S. aureus* (40 bacteria per PMN) or after stimulation with 25 ng/ml phorbol myristate acetate (PMA). Chemiluminescence produced by *S. aureus* stimulated PMN was inhibited with increasing concentrations rosmarinic acid (from 1 to 500 μM). A concentration of 500 μM rosmarinic acid decreased the peak chemiluminescence response to $54 \pm$ (SD) 10% of control values obtained in the absence of rosmarinic acid. After stimulation with PMA also a strong inhibition of the peak chemiluminescence to $20 \pm 12\%$ of control values was observed, related to the concentration of rosmarinic acid. This inhibitory effect of rosmarinic acid on the PMN chemiluminescence response was reversible. Washing after preincubation abolished the effect.

Oxygen consumption during stimulation of the metabolic burst with PMA was measured with a Gilson O_2-electrode in a microvessel [4]. After equilibration of the PMN suspension with the air, the cells were triggered with 25 ng/ml PMA and the amount of oxygen consumed was measured. During 10 min, no differences were observed regardless the absence or presence of rosmarinic acid (100 or 500 μM).

Production of hydrogen peroxide (H_2O_2), one of the important oxygen products of stimulated PMN was quantitated by the H_2O_2-mediated and horseradish peroxidase-dependent oxidation of phenol red using a standard curve of H_2O_2 concentrations (1–100 μM) [5]. In the control experiments, the amount of measurable H_2O_2 after 30 min of stimulation with 25 ng/ml PMA was 21 $\mu M/10^7$ PMN. This decreased to 6 $\mu M/10^7$ PMN in the presence of 500 μM rosmarinic acid. A similar inhibition curve was observed when rosmarinic acid was directly mixed with a constant concentration of H_2O_2 (32.5 μM). In the presence of 500 μM rosmarinic acid only 1.4 μM reactive H_2O_2 could be measured. Direct measurement of superoxide anion (O_2^-) production with the reduction of ferricytochrome c method was not useful because rosmarinic acid interfered directly with ferricytochrome c.

Another important oxygen-dependent function of PMN is their ability to lyse extracellular target cells. As a model of this extracellular cytolysis we used the PMA (10 ng/ml) stimulated destruction of ^{51}Cr-labeled autologous erythrocytes [6]. In the presence of increasing concentrations of rosmarinic acid (1–500 μM), the percentage cytotoxicity after 3 h at 37 °C decreased to 8.1 ± 2.7% of control values. Preincubation of the PMN with rosmarinic acid for 30 min, followed by washing, resulted in normal cytolysis of the labeled erythrocytes.

Discussion

Because it is shown that activated PMN may be responsible for tissue injury during inflammatory reactions, it is of special interest to use anti-inflammatory agents that are non-toxic and do not depress important parts of the host defense machinery. Our results indicate that rosmarinic acid does not affect PMN functions important in host defense against microorganisms in vitro. Chemotaxis, phagocytosis and enzymatic intracellular killing of ingested bacteria were completely normal in the presence of rosmarinic acid. Our results do not allow conclusions with regard to the intracellular oxydative killing mechanisms of the PMN, and it is also unknown where rosmarinic acid can enter the intact PMN. On the other hand, rosmarinic acid was shown to selectively inhibit effects of the extracellular released toxic oxygen products. The luminol-enhanced chemiluminescence, which involves both O_2^- and myeloperoxidase, was inhibited by rosmarinic acid after stimulation with either

PMA or opsonized bacteria. Oxygen-dependent cytolysis of erythrocytes by PMA-stimulated PMN was also reversibly inhibited in the presence of rosmarinic acid. Because oxygen consumption was normal, it seems likely that rosmarinic acid acts as a scavenger of the produced reactive oxygen products. The precise nature of the target species is not clear, but a possible role as H_2O_2 scavenger is presented through the decrease in measurable H_2O_2. Our results provide an explanatory basis for the experimental results, obtained with rosmarinic acid in a rabbit model of multiple organ inflammation, elsewhere described in this issue [2].

References

1   Babior, B. M.: Oxygen-dependent microbial killing by phagocytes, New Engl. J. Med. *298:* 659–668, 721–725 (1978).
2   Nuytinck. J. K. S.; Goris, R. J. A.; Kalter, E. S.; Schillings, P. H. M.: Oxygen free radicals and microvascular injury in a rabbit model for ARDS (this volume).
3   Verhoef, J.; Peterson, P.; Quie, P.: Kinetics of staphylococcal opsonization, attachment, ingestion and killing by human polymorphonuclear leukocytes: a quantitative assay using ³H-thymidine-labeled bacteria. J. immunol. Methods *14:* 303–311 (1977).
4   Henricks. P,; Tol, M. van der; Thijssen, R.; Asbeck, B. van; Verhoef, J.: *Escherichia coli* lipopolysaccharides diminish and enhance cell function of human polymorphonuclear leukocytes. Infect. Immunity *41:* 294–301 (1983).
5   Pick, E.; Keisari, Y.: A simple colorimetric method for the measurement of hydrogen peroxide produced by cells in culture J. immunol. Methods *38:* 161–170 (1980).
6   Weiss, S.; LoBuglio, A.: An oxygen-dependent mechanism of neutrophil-mediated cytotoxicity. Blood *55:* 1020–1024 (1980).

K. P. M. van Kessel, MD, Laboratory of Microbiology, Catharijnesingel 59, NL-3511 GG Utrecht (The Netherlands)

Novelli, Ursini (eds.), Oxygen Free Radicals in Shock. Int. Workshop,
Florence 1985, pp. 83–86 (Karger, Basel 1986)

Neutrophil-Dependent and Independent Increases in Endothelial Permeability[1]

Asrar B. Malik

Department of Physiology, Albany Medical College, Albany, N.Y., USA

An increase in endothelial permeability to protein results in increases in transendothelial fluid and protein fluxes. Tissue edema occurs from the increased movement of water and proteins across the endothelial barrier. An interaction between neutrophils and the endothelium (perhaps by a receptor-mediated event) has been suggested to be an important factor in contributing to endothelial injury and increased transendothelial permeability to protein [1, 2]. Neutrophil activation resulting in release of free oxygen radicals and granular products of neutrophils (i.e. proteases) may be the mediators of endothelial injury [1, 2]. It is unclear, however, whether the oxygen-derived free radicals or neutrophil proteases are the primary mediators of endothelial injury after neutrophil activation [3].

In the present study, an endothelial monolayer was grown on gelatinized nucleopore filters (0.8μm diameter) [4, 5]. This system was used to assess transendothelial albumin flux and to examine interactions between the endothelium and neutrophils. Bovine pulmonary endothelial cells were grown to confluence on the filters which were then mounted in a chamber having upper and lower compartments. The fluid levels of two chambers were the same, thus diffusion was the primary factor governing albumin flux across the endothelial monolayer. The permeability of ^{125}I-albumin, therefore, was a measure of macromolecule transport. Increased endothelial permeability was reflected by an increase in ^{125}I-albumin clearance.

[1] Supported by Grant HL-32418 from the National Institutes of Health.

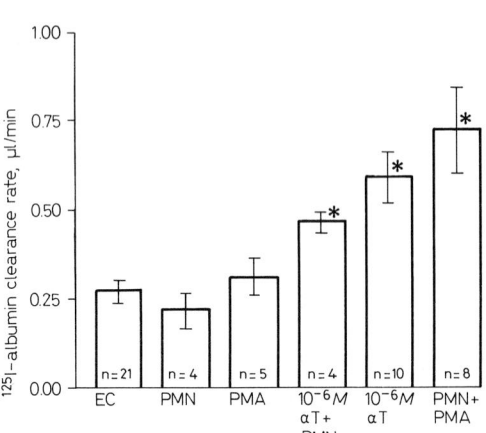

Fig. 1. Alterations in endothelial clearance rate of ^{125}I-albumin across normal endothelial (EC) + neutrophils (PMN), EC + phorbol myristate acetate (PMA) at concentration of $10^{-8}M$, EC + α-thrombin (αT) the native enzyme + PMN, EC + αT, and EC + PMN + PMA. Bars indicate standard errors and * indicate difference from the respective control ($p < 0.05$).

Albumin permeability across the endothelial monolayer is normally restricted, i.e. less than 1–2% of albumin is transferred within an hour from the upper to the lower chamber [5, 6]. The values for albumin clearance are indicated in figure 1. The layering of unstimulated sheep neutrophils slightly decreased albumin permeability. However, the addition of the neutrophil-stimulating agent, phorbol myristate acetate (PMA), increased albumin clearance, indicating that neutrophil activation results in an increase in endothelial permeability. PMA did not increase permeability without neutrophils. The increased albumin clearance was dependent upon generation of superoxide anions because superoxide dismutase (the enzyme that reduces superoxide concentration) prevented the PMA-induced increase in endothelial albumin permeability (fig. 2), indicating that PMA-induced neutrophil activation mediates the increased permeability by generation of superoxide anions.

Some agents may also have a direct effect on endothelial permeability independent of neutrophil activation. Thrombin is one such agent which was shown to increase transendothelial permeability in a concentration-dependent manner (fig. 3). The response was not enhanced by addition of neutrophils to the endothelium (fig. 1); therefore, this effect of thrombin was independent of neutrophil activation. Thrombin directly

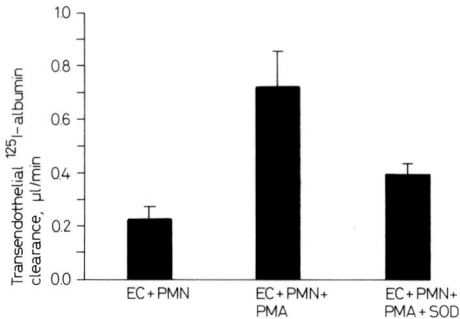

Fig. 2. Effects of superoxide dismutase (SOD) on the PMN-dependent increase in endothelial permeability induced by PMN.

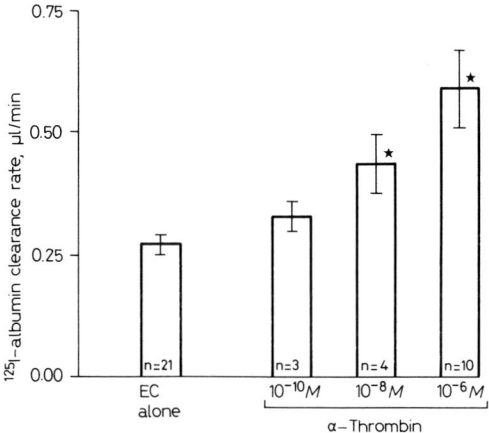

Fig. 3. Concentration-dependent increase in endothelial permeability induced by α-thrombin independent of neutrophils.

induces an endothelial permeability-increase. There was no morphological evidence of cell injury or cell lysis because lactic dehydrogenase concentration was unchanged from baseline after addition of thrombin to the endothelial monolayer. The effect of thrombin may be due to endothelial contraction due to cytoskeletal changes, and an alteration in the endothelial configuration allowing greater albumin permeability through the interendothelial junctions.

In summary, the present results indicate that endothelial injury is not a process dependent only on neutrophil activation. There are at least two mechanisms of increasing endothelial permeability: (1) a direct effect mediated by neutrophil activation and release of products from neutrophils such as superoxide anion; and (2) a direct effect mediated by agents such as thrombin which increase endothelial permeability without inducing cell injury.

References

1   Malik, A.: Pulmonary microembolism. Physiol. Rev. *64*: 1114–1207 (1983).
2   Brigham, K. L.; Meyrick, B.: Interactions of granulocytes with the lung. Circulation Res. *54*: 623–635 (1984).
3   Harlan, J. M.: Leukocyte-endothelial cell interactions. Blood *65*: 513–525 (1985).
4   Taylor, R. F.; Pine, T. H.; Schwartz, S. M.; Jale, D. C.: Neutrophil-endothelial interactions on endothelial monolayers grown on micropore filters. J. clin. Invest. *67*: 584–587 (1981).
5   Garcia, J. G. N.; Siflinger-Birnboim, A.; Bizios, R.; Del Vecchio, P.; Fenton, J. W.; Malik, A. B.: Thrombin-induced increases in albumin transport across cultured endothelial monolayers (submitted).
6   Shasby, D. M.; Shasby, S. S.; Peach, M. J.: Granulocytes and phorbol myristate increase permeability across cultured endothelial monolayers and isolated perfused lungs. Am. Rev. resp. Dis. *127*: 72–76 (1983).

Dr. Asrar B. Malik, Department of Physiology, Albany Medical College of Union University, 47 New Scotland Avenue, Albany, NY 12208 (USA)

Novelli, Ursini (eds.), Oxygen Free Radicals in Shock. Int. Workshop,
Florence 1985, pp. 87–93 (Karger, Basel 1986)

Superoxide Generation by Granulocytes during Superior Mesenteric Artery Occlusion Shock in Rabbits

*Gian Paolo Novelli[a], Paola Livi[a], Maria Luisa Ghinassi[a], Loris Lisi[a],
Sandra Brunelleschi[b], Roberto Fantozzi[b]*

[a] Institute of Anaesthesiology and Intensive Care; [b] Department of
Pharmacology, University of Florence, Italy

Introduction

The role of oxygen free radicals in the pathogenesis of shock seems
to be well supported, although the direct proof has not been achieved.

The main cellular source of oxygen radicals are activated granulocytes (PMN) that may generate them independently of phagocytosis
[*Goldstein* et al., 1975].

PMN activation has been reported to be present in every type of
shock and to be prevented by steroids [*Novelli* et al., to be published;
Redl et al., 1984].

Here reported experiments were directed to study the superoxide
(O_2^-) generation from PMN of rabbits submitted to superior mesenteric
artery occlusion (SMAO); blood was taken from systemic circulation and
from mesenteric vein.

Because steroids are known to prevent complement activation and
PMN aggregation, other experiments were performed to observe the ability of these drugs to modify the O_2^- generation from granulocytes.

Material and Methods

Experiments were performed on female New Zealand rabbits (2.5–3 kg) fasted
24 h before experiment.

Light general anesthesia was obtained with ketamine (25 mg·kg^{-1} i.m.); atropine
sulfate (0.002 mg·kg^{-1} i.m.) was also given.

A polyethylene catheter (20 G) was introduced in the femoral vein and connected to a slow infusion of Ringer lactate solution. A polyethylene catheter (20 G) was introduced in the femoral artery and placed in the abdominal aorta; it was connected to a mercury manometer through a three-way manifold to take blood for acid-base and O_2^- generation measurements.

A median laparotomy was performed and a tourniquet was passed around the superior mesenteric artery (SMA) so to make possible to occlude it through the abdominal wall closed to prevent water and heat dispersion. A heating lamp was used during the whole experiment.

To provoke shock, the tourniquet was tightened and after 30 min it was slowly loosened; according to previous experiments this period of SMAO is followed by the death of all animals within 3 h.

To measure blood gases and O_2^- generation, 4 ml of arterial blood was withdrawn with heparin and replaced with a similar volume of Ringer solution through the venous line. Blood gases and acid-base status were measured with a micromethod (IL 1312, Milan, Italy).

Superoxide production by whole blood (as superoxide dismutase-sensitive cytochrome c reduction) was measured according to *Bellavite* et al. [1983] in the absence and in the presence of N-formylmethionyl-leucyl-phenylalanine (FMLP: Serva) or A 23187 (Calbiochem).

Aliquots ($\simeq 10^6$ neutrophils) of blood samples withdrawn from femoral artery or mesenteric vein were treated at 37 °C for 5 min with cytochalasin B (5 g/ml: Aldrich). Stimuli ($10^{-7} M$ FMLP; $10^{-5} M$ A23187) were added and the incubation was allowed to proceed for further 5 min. Superoxide production from unstimulated samples was monitored after the same incubation time following cytochalasin B treatment. Absorbance values were measured at 550–468 nm.

Experiments

(1) After 30 min of stabilization, the first blood sample was taken (T_0) and SMA was occluded. Two other samples were obtained 5 min (T_5) and 25 min (T_{25}) after SMAO. After 30 min of occlusion, the tourniquet was loosened and the mesenteric bed reperfused. After 10 min from the reperfusion (T_{40}) another sample of systemic arterial blood was obtained. Immediately after this (T_{45}), a sample of mesenteric venous blood was obtained by direct puncture. Finally, the last arterial sample was taken 30 min after mesenteric reperfusion (T_{60}).

(2) Other experiments were performed according to the same sequence but methylprednisolone (30 mg·kg^{-1}) was injected in the venous line before SMAO.

Results

The shock model used appeared to be lethal in a short time and in the present experiment it was worsened by the blood withdrawals, although a certain degree of correction was attempted by infusion of polysaline solution.

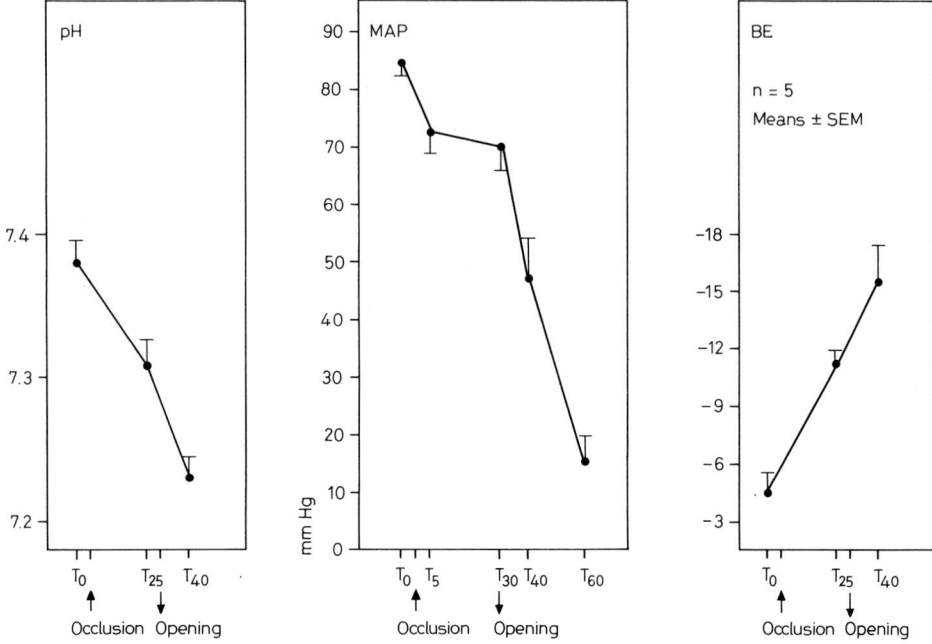

Fig. 1. Acid-base parameters and mean arterial pressure (mean values ± SE) of rabbits submitted to SMAO shock.

Uncompensated shock in rabbits submitted to SMAO initiates following opening the tourniquet and reperfusion of the mesenteric vascular bed. As reported in figure 1, acid-base status and mean arterial pressure were not significantly affected during the period of surgery and of SMAO. However, immediately after release of the tourniquet and reperfusion of the mesenteric vascular bed, all the examined parameters deteriorated very rapidly until death. Therefore, shock appeared not to be provoked by the occlusion per se but to be dependent on the reperfusion of the ischemic mesenteric bed.

Figure 2 summarizes the amounts of O_2^- spontaneously generated or produced after stimulation with FMLP or with A23187. Artery occlusion was followed by an increased generation; however, generation markedly increased after mesenteric reperfusion (T_{40}). In this moment, a correspondent decrease in evoked superoxide generation was measured. At the same time (fig. 2b) spontaneous O_2^- generation by venous mesenteric blood was high and stimulation was quite ineffective.

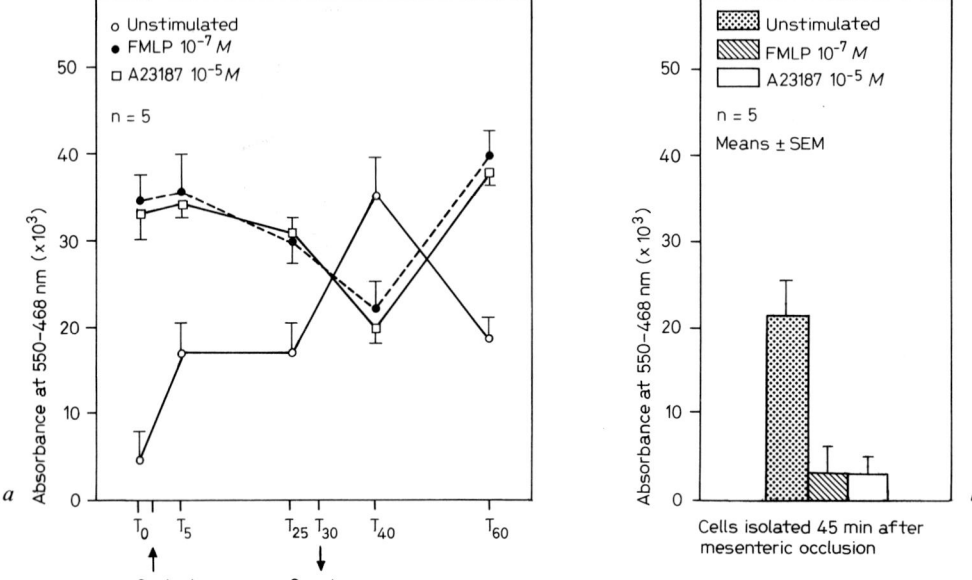

Fig. 2. Superoxide generation (spontaneous and after two stimuli) in whole blood of rabbits submitted to SMAO shock. *a* Data of systemic arterial blood. *b* Data of mesenteric venous blood.

Figure 3 summarizes acid-base status and blood pressure observed in rabbits submitted to SMAO as before but receiving methylprednisolone. All parameters were less deteriorated than in previous experiments.

Figure 4 shows that methylprednisolone injected before SMAO prevented every increase in O_2^- generation both after occlusion and after reperfusion of the mesenteric bed. Challenged cells, however, produced significant amounts of superoxide. In blood taken from the mesenteric vein (figure 4b) spontaneous generation was absent but the stimuli were very effective.

Fig. 3. Acid-base parameters and mean arterial pressure (mean values ± SE) of rabbits submitted to SMAO shock after a bolus of methylprednisolone (30 mg·kg^{-1} i. v.). Differences in respect to non-treated animals were evident.

Fig. 4. Superoxide generation (spontaneous and after two stimuli) in whole blood of rabbits submitted to SMAO shock after a single bolus of methylprednisolone (30 mg·kg^{-1} i. v.). *a* Data of systemic arterial blood. *b* Data of mesenteric venous blood. Compared with figure 2.

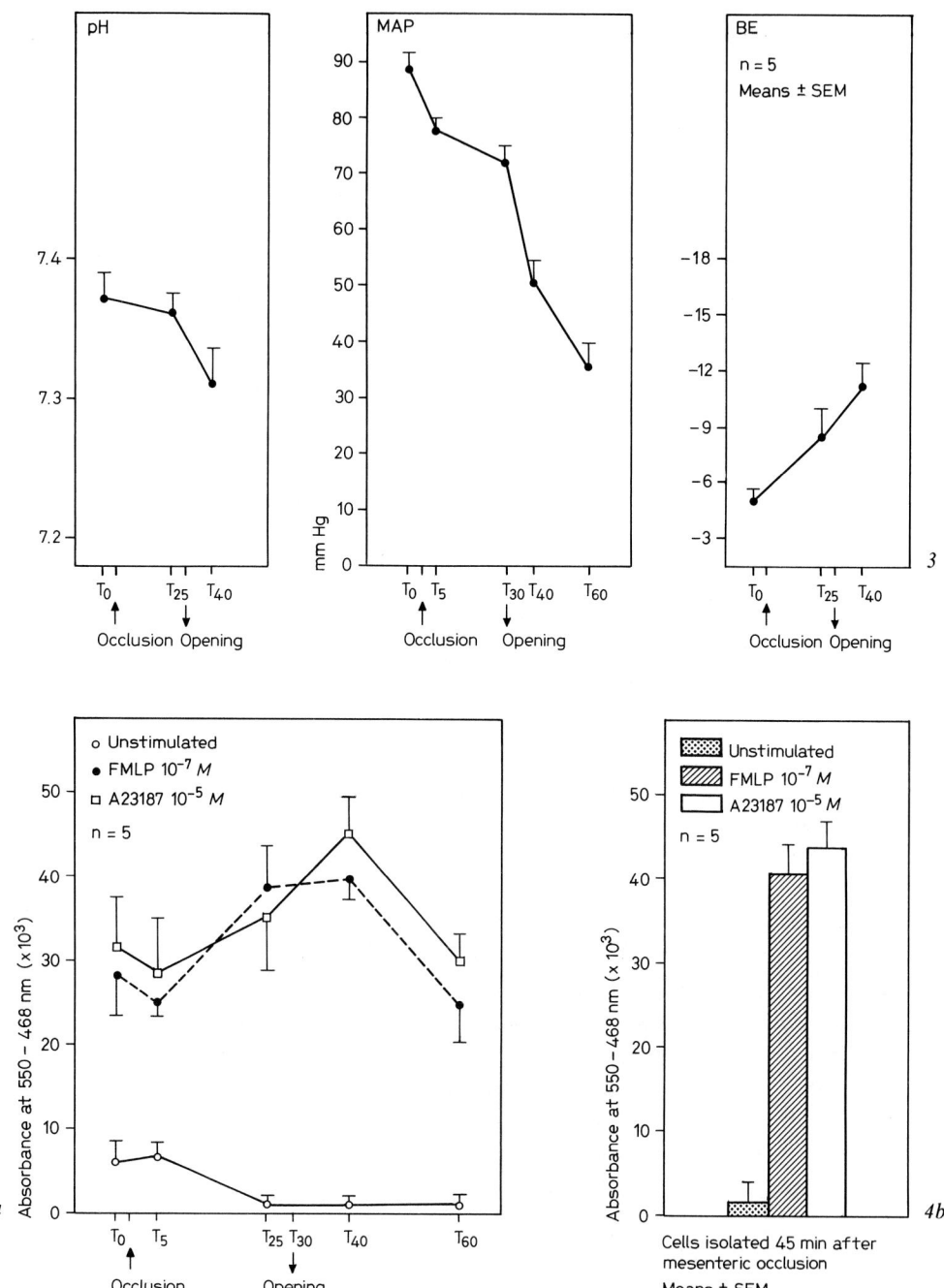

Discussion

Here reported experiments indicate that during lethal SMAO shock in rabbits there was an increased production of O_2^- in whole blood. Anesthesia, fasting and surgery might be able to provoke a certain degree of PMN activation. In fact, in the T_0 blood sample there was a spontaneous production of superoxide. At T_5, when SMA was occluded, the spontaneous O_2^- production increased and remained constant during the whole time before the opening of the artery. This PMN activation was absent in methylprednisolone-pretreated animals. Therefore, it may be indicative of a free-radical production in response to trauma and to ischemia consequent to diffuse reflex vasoconstriction.

O_2^- generation at T_{25} was very low in methylprednisolone-pretreated animals and it remained low after reperfusion of the mesenteric bed and mixing of its blood into the systemic circulation. During the whole experiment, PMN were very responsive to stimulation with both stimuli.

In experiments without steroids, the reperfusion of the mesenteric bed was followed by a sudden increase of O_2^- generation with a parallel decreased response to stimuli. The same was observed in venous blood taken from mesenteric bed.

Therefore, a massive PMN activation might happen in the mesenteric occluded bed; systemic activation may follow through a cascade sequence involving the complement system.

In experiments reported here, methylprednisolone was effective in preventing PMN activation without affecting their responsiveness to exogenous stimuli. Phagocytosing PMN were observed to generate diminished levels of O_2^- when exposed in vitro to steroids [*Fuenfer* et al., 1979; *Levine* et al., 1981]. Human leukocytes also were reported to decrease O_2^- production in response to concavalin A after exposition to steroids both in vivo and in vitro [*Nelson* et al., 1981]; this effect was related to a variation of cell phospholipid component.

In present experiments, the ability of PMN to produce O_2^- in response to exogenous stimulations was unaffected by steroids and spontaneous generation only was depressed. Therefore, the activity of methylprednisolone on O_2^- production during and after SMAO may be related to an inhibitory effect on complement system [*Hammerschmidt* et al., 1979; *Imai* et al.; 1982].

References

Bellavite, P.; Dri, P.; Della Bianca, V.; Serra, M. C.: The measurement of superoxide anion production by granulocytes in whole blood. A clinical test for the evaluation of phagocyte function and serum opsonic capacity. Eur. J. Clin. Invest. *13:* 363–368 (1983).

Fuenfer, M. M.; Carr, E. A; Polk, H. C.: The effect of hydrocortisone on superoxide production by leukocytes. J. surg. Res. *27:* 29–35 (1979).

Goldstein, I. M.; Roos, D.; Kaplan, H. B.; Weissman, G.: Complement and immunoglobulins stimulate superoxide production by human leukocytes independently of phagocytosis. J. clin. Invest. *56:* 1155–1163 (1975).

Hammerschmidt, D. E.; White, J. G.; Craddock, P. R.; Jacob, H. S.: Corticosteroids inhibit complement-induced granulocyte aggregation. A possible mechanism for their efficacy in shock states. J. clin. Invest. *63:* 798–803 (1979).

Imai, T.; Sato, T.; Fujita, T.: Inhibitory effect of glucocorticoids on complement activation induced by lipopolysaccharide. Circul. Shock *9:* 55–62 (1982).

Levine, P. H.; Hardin, J. C.; Scoon, K. L.; Krinsky, N. I.: Effect of corticosteroids on the production of superoxide and hydrogen peroxide and the appearance of chemiluminescence by phagocytosing polymorphonuclear leukocytes. Inflammation *5:* 19–27 (1981).

Novelli, G. P.; De Gaudio, A. R.; Bianchi, A.; Bagatti, S.: Polymorphonucleate aggregation and oxygen-derived free radicals during endotoxin and trauma shock. Acta anaesth. scand. (accepted for publication).

Nelson, D. H.; Wenhold, A. R.; Murray, D. K.: Corticosteroid induced simultaneous changes in leukocyte phospholipids and superoxide anion production. J. Steroid Biochem. *14:* 321–325 (1981).

Redl, H.; Schlag, G.; Hammerschmidt, D. E.: Quantitative assessment of leukostasis in experimental hypovolemic-traumatic shock. Acta chir. scand. *150:* 113–117 (1984).

Prof. G. P. Novelli, Institute of Anaesthesiology and Intensive Care, University of Florence, Policlinico di Careggi, V. le Morgagni, I-50134 Florence (Italy)

Novelli, Ursini (eds.), Oxygen Free Radicals in Shock. Int. Workshop,
Florence 1985, pp.94–108 (Karger, Basel 1986)

Oxygen Radicals in Hypovolemic-Traumatic Shock[1]

Günther Schlag, Heinz Redl[2]

Ludwig Boltzmann Institute of Experimental Traumatology, Vienna, Austria

Introduction

In hypovolemic-traumatic shock there are three possible sources of
oxygen radicals – phagocytes (PMN, macrophages), ischemic peripheral
tissue (xanthine-oxidase action) and by-products of the 'arachidonic-cas-
cade'.

Normally the oxygen radicals cause hardly any tissue problems due
to their very short half-life and to the balance between free radical generation and neutralization by endogenous defense systems which protect
the integrity of cells and tissues [1].

In shock, an insufficient defense system and excessive free oxygen
radical generation may result in severe cell damage. Damage may be
either direct, or indirect via inactivation of proteinase inhibitors, and
thus may give rise to proteolytic injury.

Our group has shown that post-traumatic hypovolemic shock produces pulmonary changes which predominantly involve cellular and subcellular areas. These morphological changes are only demonstrated on
the ultrastructural level and can be seen immediately after injury [2].
Therefore, this phase is called 'lung in shock', or referring to the entire
organism, 'the organ in shock' (liver, pancreas, intestine, kidney, heart,
brain, skeletal muscles).

[1] This study was supported by a grant from the 'Nationalbankfonds'.

[2] We thank Mrs. *E. Paul, A. Schiesser, Ch. Vogl, C. Wilfing*, Dr. *A. Fried*, Dr. *M.
Turnher*, Dipl. Ing. *P. Weimerskirch, W. Junger, C. Lieners* for their excellent technical
assistance and Mrs. *B. Pavlis* for preparing the manuscript.

The endothelial cells of the capillaries, especially in the lung, are among the targets of the free oxygen radicals. The most prominent ultrastructural alterations of the endothelial cells are a variable swelling to the point of necrobiosis in localized areas of a capillary. This endothelial cell damage may be caused by several mediators and toxic products. In addition to toxic oxygen radical release, other sources are possible such as release of neutrophil proteinases (i.e. elastase). In the lung, additional hypoxic damage may be present due to ventilation/perfusion inequalities during the shock period.

Oxygen radicals may initiate lipid peroxidation and thus directly damage the cell membrane, which contains a large amount of polyunsaturated fatty acids [1]. The change in membrane lipids alters the functional integrity of the membrane and increases permeability [3]. This may be the cause of incipient interstitial edema of the lung in shock, which is already manifest during the shock phase and particularly prominent after reinfusion and transfusion.

The morphological hallmark of practically any type of shock is leukostasis of the lung. Lung biopsies of patients in shock have indicated that severe leukostasis is always associated with PMN degranulation [4].

The sequestration of stimulated leukocytes (e.g. by complement activation products, C3a, C5a) in the microvasculature of the lung leads to a release of toxic oxygen radicals, proteolytic enzymes (i.e. elastase, collagenase) and inflammatory mediators (i.e. arachidonic cascade-derived products).

The importance of PMN stimulation and accumulation is corroborated by our recent study with *Goris* [5], where the levels of circulating granulocyte elastase proteinase inhibitor complex correlated well with injury severity. Elastase appeared to be the best predictor of adult respiratory distress syndrome (ARDS) upon admission and after 24 h. This seems to prove the important role of activated PMN in connection with ARDS and is in accordance with our morphological findings in lung biopsies of shocked patients.

Ischemic tissue is another source of oxygen-derived free radical release. Hemorrhagic shock may be viewed as 'whole body ischemia' [6], since tissues are insufficiently perfused and, as a consequence, ATP levels are decreased in various organs such as liver, pancreas, intestinal and skeletal muscles. This, in turn, increases the hypoxanthine level in the plasma.

Restoration of tissue perfusion and oxygenation with fluid treatment may activate xanthinoxidase with resultant production of O_2^-, and this may further lead to reperfusion injury of the capillary wall with localized endothelial swelling as an expression of cell damage.

In a previous study we compared skeletal muscle biopsies with a lung biopsy acquired from the same patient in shock. We found ultrastructural endothelial changes in the microvascular system of the lung and skeletal musculature, however, with rare evidence of leukocytes in the latter [7].

In our studies, we investigated (1) the activation of the complement system with the occurrence of leukostasis in the lung and liver, (2) the levels of circulating malondialdehyde-like material, (3) the ischemia in muscle tissue by pH, adenine nucleotide content and nucleoside plasma level measurements and, finally, (4) the resulting permeability increase by estimating the transcapillary escape rate (TER) and the extravascular lung water (EVLW). Since all these parameters could not be obtained simultaneously from each animal, two series with an identical shock protocol were performed.

Materials and Methods

Anesthesia was induced by pentobarbital 30 mg/kg and maintained with piritramide 0.5 mg/kg/h, relaxation with pancuronium 0.1 mg/kg/h. Ventilation was provided by a constant volume ventilator (15 ml/kg tidal volume at a rate of 16/min). Arterial pCO_2 was kept slightly below the normal range (35 ± 4 mm Hg) by adjustment of minute ventilation.

The animals were instrumentated for measurement of the aortic, pulmonary artery and right atrial pressure. Cardiac output was determined by thermodilution, heart rate by electrocardiography. To measure the muscle pH, an Ingold 406-M3 electrode was placed in the oblique external abdominal muscle. In a total of 26 anesthetized mongrel dogs (21.3–27.5 kg BW) both femoral bones were closed fractured and 100 blows were administered to one thigh. Subsequently, the animals were bled to a mean arterial blood pressure of 40 mm Hg. 30–45 min were required to obtain a stable blood pressure of 40 ± 5 mm Hg. Throughout the experiment, the animals were maintained at this level of hypotension (for 2 h). They were initially reinfused within 45 min with their own shed blood (2,000 IU heparin/500 ml blood) plus Ringer albumin solution (2.5%) up to a cardiac output of 120% of the control value without exceeding a pulmonary artery pressure of 25 mm Hg. A further discontinuous fluid replacement was done until the end of the experiment using the same criteria.

Experimental Procedure

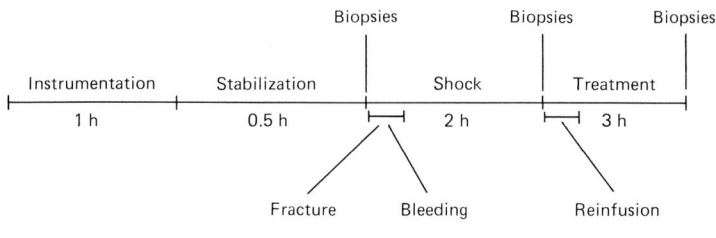

For lung water determination and histological examination, lung tissue was taken from the left lower lobe. Wedge biopsies were taken from the liver. Muscle tissue was sampled for adenine nucleotides from the lateral part of the thigh muscles (M. vastus lateralis) at the end of the shock and 3 h after reperfusion, on the side which was not injured by blowing. Heart muscle tissue was taken from the left ventricle.

Light and electron microscopic studies were done with our previously reported protocol [8]. Total complement level was determined as CH_{50} according to *Kabat and Meyer* [9].

The transcapillary albumin escape rate (TER, i.e. % intravascular activity loss/h) was determined by the method reported by *Parving and Gyntelberg* [10]. For the first determination we injected ^{125}I-albumin (20 µCi), while ^{131}I-albumin (0.4 mCi) was used for the second determination. These doses (19 ml) were administered by barbotage (axillary vein) in heparinized (1,000 U) Ringer's lactate containing 0.1 % albumin. Blood radioactivity was measured by a Beckman Instruments Biogamma counter. A selective window (0.50 to 1.00 keV) was used to discriminate ^{131}I from ^{125}I.

Extravascular lung water (EVLW) was measured gravimetrically as previously described [11]. Protein and lactate were measured with kits from Boehringer, Mannheim, β-glucuronidase activity was analyzed according to *Szasz* [12].

The determination of thiobarbituric acid-reactive (TBA) material calculated as malondialdehyde (MDA), and the separation of nucleotides (ATP/ADP) and nucleosides (hypoxanthin) by HPLC is described by *Redl* et al. in this volume, except that for plasma nucleosides blood was withdrawn in dipyridamole.

The differences between groups were evaluated with the unpaired t-test. Differences between samples were analyzed by paired t-test. The significance levels were defined as at least 5 % probability of error (p < 0.05).

Results

Morphologically granulocytes were again found to accumulate in the lungs of the traumatized animals at the end of experiment (fig. 1). Granulocyte accumulation was, in part, associated with endothelial cell changes. Leucostasis was, as a rule, accompanied by degranulation. In-

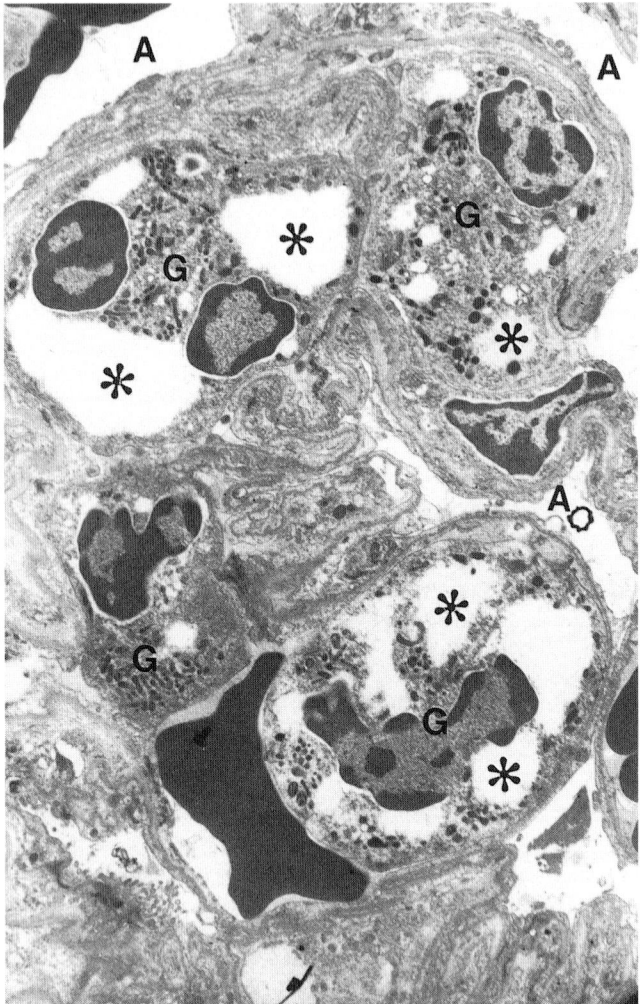

Fig. 1. Fulminant leukostasis in the lung with degranulation (*) of neutrophils (G). A = Alveolus. EM. × 2,900.

Fig. 2. Muscle specimen with capillary. Swollen endothelium (*) is easily detected. E = Erythrocytes. EM. × 3,950.

Fig. 3. Capillary of left ventricular heart muscle with typical focal endothelial swelling (*), which is limited by cell border (junctions →). The capillary lumen is filled with an erythrocyte (E). EM. × 2,450.

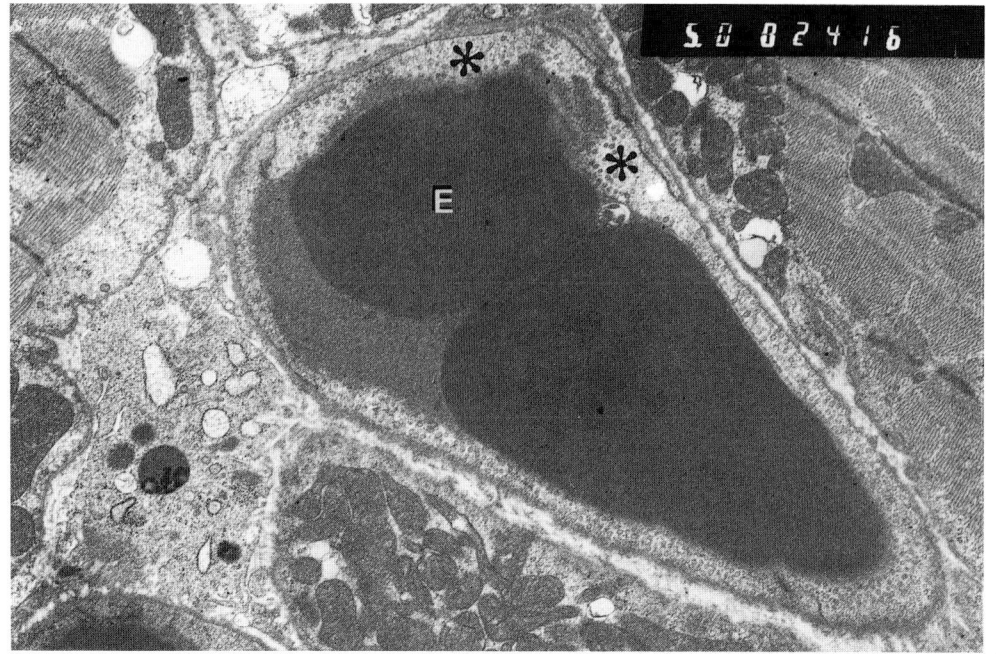

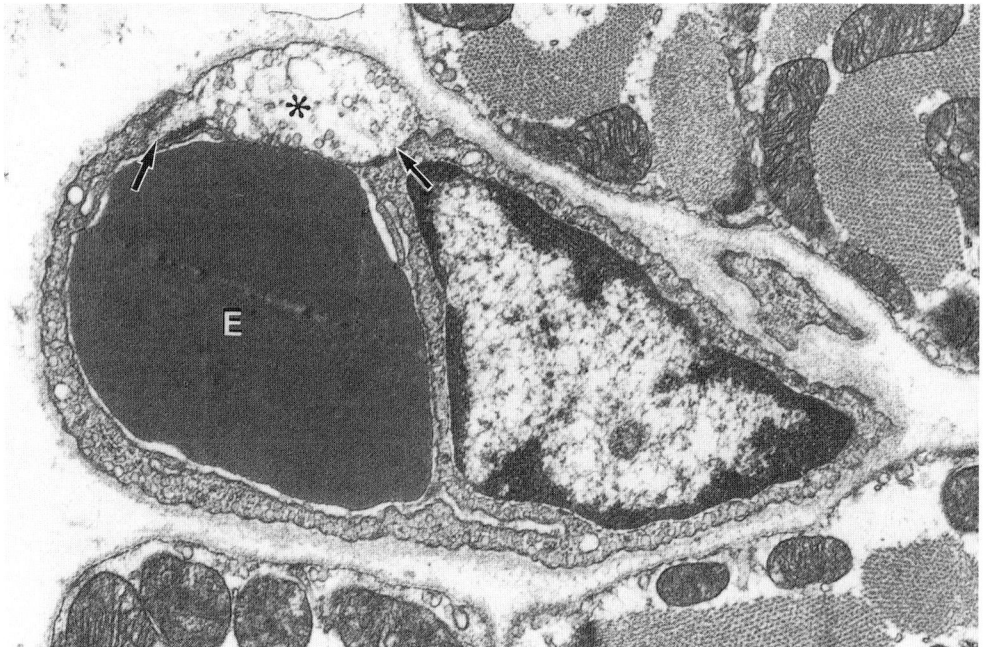

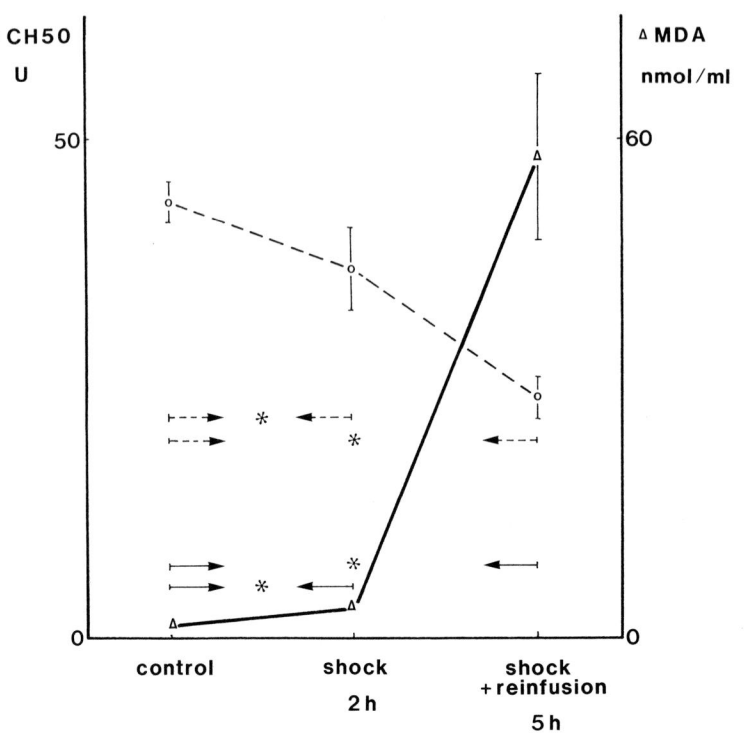

Fig. 4. Significant increase of malondialdehyde (Δ MDA) like material in plasma. Total complement (o CH_{50}) plasma level is significantly reduced. $* = p < 0.05$.

cipient interstitial pulmonary edema tended to occur, especially at sites primarily not involved in pulmonary gas exchange.

In muscle tissue, capillary endothelial swelling without accumulation of PMN was seen (fig. 2). Further changes could not be detected. Similar swelling could be shown in heart muscle tissue obtained from the left ventricle (fig. 3).

At the end of the experiment, liver biopsies contained swollen Kupffer cells with phagotized degranulated or even fragmented PMN and hemolysed erythrocytes. Similar to the other tissues, sinusoidal endothelial swelling was seen. In addition. widening of Disse's spaces was encountered.

Muscle tissue showed diminished ATP content in shock, which was significantly ($p < 0.05$) increased after reinfusion (ATP/ADP – table II).

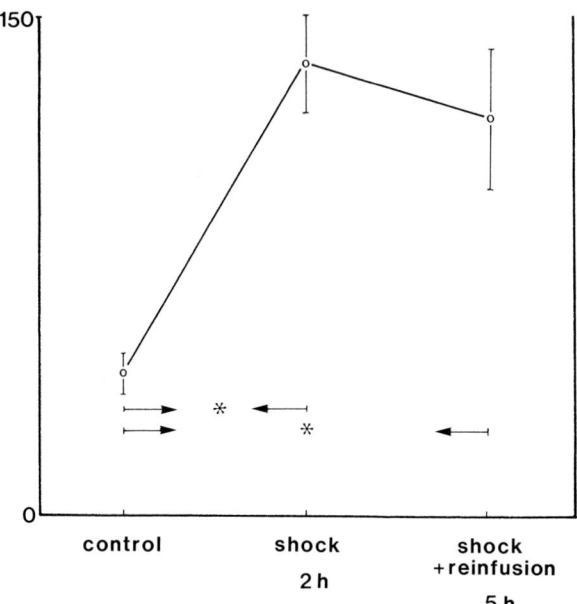

Fig. 5. β-Glucuronidase is significantly increased during the shock period. * = p < 0.05.

At the same time, nucleoside content showed a nearly significant (p = 0.056) decrease after restoration of muscle blood flow (table II). Similarly, the plasma HXA+XA content was diminished compared with the end of shock, where an increase of 13-fold although not statistically significant (p = 0.08) was seen compared with the control plasma (table I). There was a parallel significant decrease (p < 0.001) of muscle tissue pH and a significantly (p < 0.001) increased lactate A/V difference measured between the femoral artery and vein.

TBA reactive material showed a significant (p < 0.01) elevation from control, both at the end of the shock period and after reinfusion (p < .0.001) (fig. 4). During the same time period activation of complement occurred with a concomitant significant (p < 0.001) decrease in CH_{50} (fig. 4).

Transcapillary Escape Rate

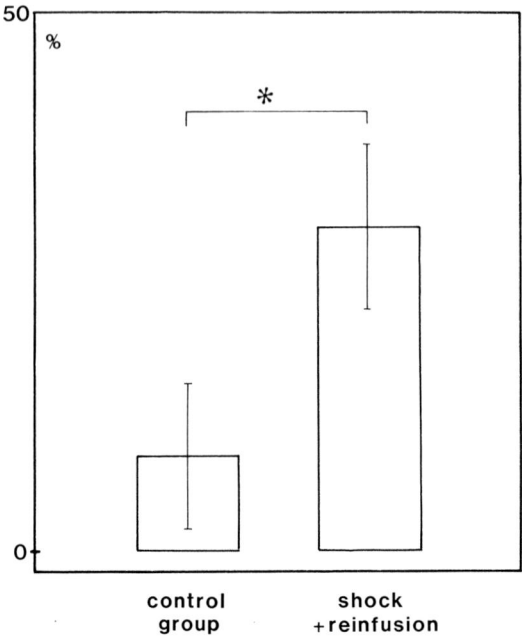

Fig. 6. Significant increase in whole body permeability measured as transcapillary escape rate (= % loss of labeled albumin/h). * = p < 0.05.

The time course was different for the release of β-glucuronidase, where a significant (p < 0.01) increase was seen at the end of shock followed by a more or less similar level after retransfusion (fig. 5).

Permeability changes during the experiment led to a significant (< 0.05) increase in the TER (fig. 6) and in lung EVLW content (p < 0.001) (fig. 7).

Discussion

Endothelial cell damage in hypovolemic traumatic shock affects the lung as well as other vital organs such as liver, pancreas, intestinal tract, kidney and the skeletal musculature. Cell damage may be due to toxic O_2 radical produced by phagocytes (macrophages, granulocytes) and by

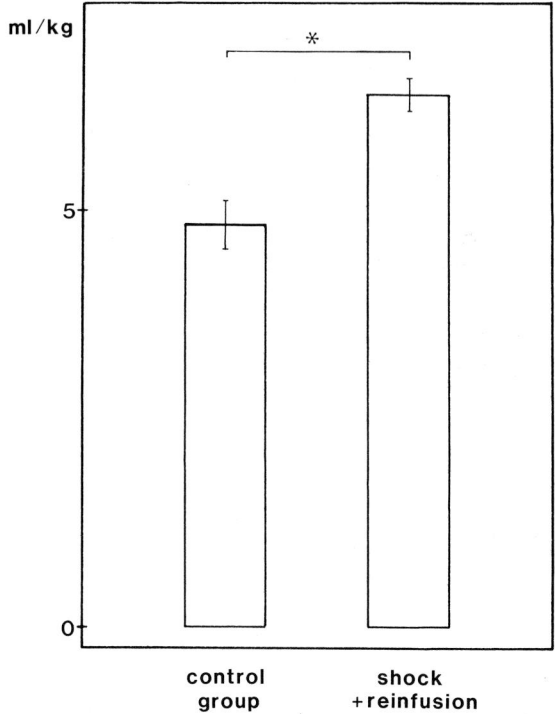

Fig. 7. Significant difference of EVLW between normal lungs (control, n = 16) and lungs after the experiment (shock+reinfusion, n = 10). *=p<0.05.

xanthinoxidase, as well as to proteinases (granulocytes) and hypoxic damage.

Since leukostasis has been confirmed both in qualitative [2, 13–20] and in quantitative [21, 22] terms, we are convinced that PMN are extensively involved in endothelial cell damage of the capillaries in the lung.

As a rule, pulmonary leukostasis is associated with degranulation of the PMN [16]. This has been confirmed in human and animal biopsies. Tissue leukostasis was accompanied by evidence of lysosomal enzyme release – e.g. significant β-glucuronidase increase (fig. 5), as reported previously [4]. Endothelial damage, therefore, may result from the joint action of oxygen radicals and lysosomal enzymes.

An elevated PMN activity in polytraumatized patients was proven by the correlation between neutrophil elastase and the degree of polytrauma. A highly significant relationship was noted between elastase concentration within 24 h following polytrauma and the prediction of ARDS. Post-traumatic ARDS thus seems to be primarily due to migration of activated PMN into the lung. As described in a number of reports [2, 23, 24], complement activation occurs during shock [25, 26] with resultant C5a release. This, in turn, causes stimulation and aggregation of PMN. In the experiment, a marked decrease of the overall complement CH_{50} was seen (48 ± 2 U $\rightarrow$ 24 ± 2 U).

Heidemann [26] was among the first to demonstrate complement activation in vitro with incubation of plasma and crushed muscle tissue. Previous morphological findings in the lung have prompted us to undertake ultrastructural investigations of the microvasculature in liver and skeletal muscles. In the liver, severe leukostasis with degranulation and endothelial swelling occurred. This corroborates other studies where vacuoles in hepatocytes, swelling of mitochondria, widening of the endoplasmatic reticulum and intracellular edema were described [27, 28].

In the liver, a significant increase of labelled neutrophil-based tissue radioactivity after shock and retransfusion [29] had been noted previously. The liver is apparently similar to the lung, in that it acts – at least partly – as a sieve for activated or damaged PMN from the peripheral circulation.

In the skeletal muscle, endothelial cell damage may have different genesis. During shock, the arterial and venous flow as well as the muscle pH decrease substantially (from 7.38 ± 0.03 to 6.80 ± 0.05) (table I) as an expression of partial ischemia. This is reflected by a high local arteriovenous lactate difference of almost 2 mmol/l (table I).

Van der Kley et al. [30] noted a decrease of about 80% in muscle pO_2 during hemorrhagic shock. Thus, one may suggest that muscular cell damage is primarily due to hypoxia. In muscle biopsies acquired during shock [29], but before reinfusion, considerably less evidence of endothelial cell damage was found. Neither during shock nor following reinfusion an increase of labelled granulocytes was noted. The source of oxygen radicals (O_2^-) during ischemia is the enzyme xanthinoxidase, since its inhibitor allopurinol prevents cell damage (e.g. mucosal lesions of the small intestine after ischemia).

Furthermore, allopurinol increased the survival rate of dogs following hemorrhagic shock to 6 times the original value [31]. In shock, the necessary substrate of xanthinoxidase, i.e. hypoxanthin, is derived from the degradation of ATP during ischemia (whole body ischemia according

Table I. Muscle pH, lactate femoral A/V difference and plasma hypoxanthin and xanthin concentration during control, shock and reinfusion period

	Tissue pH	A/V lactate difference mmol/l	Hypoxanthin (HX) + xanthin (X) nmol/ml plasma
Control (C)	7.38±0.03	–0.06±0.10	2.1± 0.8
Shock (S)	6.80±* 0.05	–2.00±* 0.32	28.4±11.3
Shock + reinfusion (R)	7.22±0.03	–0.24±0.37	8.0± 4.7

Table II. ATP/ADP ratio and total xanthin content in thigh muscle tissue in shock and after retransfusion

	Muscle tissue content	
	ATP/ADP ratio	HX+X nmol/μg DNA
Shock (S)	4.74±0.3	0.94±0.14
Reinfusion (R)	6.05±* 0.5	0.70±0.14

to *McCord* [6]). Hypoxanthin in the plasma levels rose from 2.1±0.8 mmol/l (control) to 28.4±11.3 mmol/l (table I), while the ATP/ADP ratio was 4.74 during shock as against 6.05 during reperfusion (table II).

During hypoxia and/or ischemia, no visible ultrastructural cell damage of the capillary endothelium is noted, only after reperfusion, i.e. when molecular oxygen is reintroduced to the tissue [32]. The toxic O_2 radicals may lead to lipid peroxidation in the biologic membranes, especially of the polyunsaturated fatty acids, such as arachidonic acid. Lipid peroxide radicals, lipid hydroperoxides and lipid endoperoxides as well as the degradation product MDA are found [1]. As our results indicate, MDA increases (1.4±0.1 nmol/l), significantly during shock (3.5±0.5 nmol/l), and massively after reinfusion (57.7±10.2 nmol/l) (fig. 4).

Through the reconstitution of the blood volume, tissue perfusion and oxygenation is restored and results in free radical-induced reperfusion injury [6].

In previous experiments fluid accumulation in the interstitium was not seen during the hypovolemic period and the low flow stage of shock, because radical injury is only massive after reperfusion. In fact, minor interstitial edema does not develop before the circulatory volume has been restored [3]. In addition to the free radical-induced reperfusion injury, possibly reflected in dramatic MDA elevation after retransfusion, the elevated microvascular pressure during the resuscitation/reinfusion period may contribute to fluid accumulation, e.g. of the lung (fig. 7).

In contrast, whole body permeability parameters such as TER was significantly increased already during the shock, which may reflect, at least in part, loss of tracer to the traumatized tissue and dilution by intravasation of interstitial fluid due to hypotension. Reperfusion injury in shock seems to be the reason for endothelial cell swelling of the capillaries in the myocardium, especially in the subendocardial region, where the flow in hypovolemic shock is markedly decreased [34–36]. Thus, nonperfused areas of the myocardium may be encountered, and as a consequence, a reperfusion syndrome occurs (fig. 3).

From all this rather indirect evidence we can conclude that a pathomechanism involving the oxygen-derived free radicals is indeed operative in hypovolemic traumatic shock. Therefore, scavengers of free radicals may become an important therapeutic agent in the future.

References

1   Bertrand, Y.: Oxygen-free radicals and lipid peroxidation in adult respiratory distress syndrome. Intensive Care Med. *11:* 56–60 (1985).

2   Schlag, G.; Redl, H.: Morphology of the posttraumatic human lung after traumatic injury; in Zapol, Falke, Pathophysiology and therapy of severe acute lung disease, pp. 161–183 (Dekker, New York 1985).

3   Vladimirov, Y. A.; Olenev, V.I.; Suslova, T. B.; Cheremisina, Z. P.: Lipid peroxidation in mitochondrial membrane. Adv. Lipid Res. *17:* 173–180 (1980).

4   Schlag, G.; Redl, H.: Die Leukostase in der Lunge beim hypovolämisch-traumatischen Schock. Anaesthesist *29:* 606–612 (1980).

5   Nuytinck, H. K. S.; Goris, R. J. A.; Redl. H.; Schlag, G.; Munster, P. J. J. van: Posttraumatic complications and inflammatory mediators. Archs. Surg., Chicago (submitted).

6   McCord, J. M.: Oxygen-derived free radicals in postischemic tissue injury. New Engl. J. Med. *312:* 159–163 (1985).

7   Schlag, G.; Voigt, H. W.; Schnells, G.: Vergleichende Untersuchungen der Ultrastruktur von menschlicher Lunge und Skelettmuskulatur im Schock. Anaesthesist *26:* 612–622 (1977).

8   Redl, H.; Schlag, G.; Grisold, W.; Stachelbergr, H.: Early morphological changes of the lung in shock demonstrated in the light (LM), transmission electron (TEM) and scanning electron microscopes (SEM). Scann. Electr. Microsc. *2:* 555–562 (1980).

9   Kabat, E. A.; Mayer, M. M.: Experimental Immunochemistry; 2nd ed. (Thomas, Springfield 1971).

10 Parving, H. H.; Gyntelberg, F.: Transcapillary escape rate for albumin and plasma volume in essential hypertension. Cirulation Res. *32:* 643–650 (1973).

11 Redl, H.; Schlag, G.: Permeabilitätsuntersuchungen an einem hypovolämisch-traumatischen Schockmodell bei Spontanbeatmung unter Verwendung einer Ringer- beziehungsweise Albumin-Lösung zur Volumssubstitution. Hft Unfall-heilk. *156:* 572–583 (1983).

12 Szasz, G.: Die Bestimmung der ß-Glucuronidase Aktivität im Serum mit *p*-Nitro-phenylglucuronid. Clin. Chim. Acta *15:* 275–280 (1976).

13 Schlag, G.; Regele, H.: Lungenbiopsien bei hypovolämisch-traumatischem Schock. Med. Welt *23:* 1755–1758 (1972).

14 Schlag, G.; Voigt W. H.; Schnells, G.; Glatzl, A.: Die Ultrastruktur der menschlichen Lunge im Schock. Anaesthesist *25:* 512–521 (1976).

15 Redl, H.; Schlag, G.: Das 'Lung Converting Enzyme' als möglicher Zellmarker in Verbindung mit strukturellen Veränderungen der Lunge im hypovolämisch-traumatischen Schock. Anaesthesist *29:* 552–558 (1980).

16 Schlag, G.; Voigt, W. H.; Redl, H.; Glatzl, A.: Vergleichende Morphologie des posttraumatischen Lungenversagens. Anaesth. Intensivther. Notfallmed. *15:* 315–339 (1980).

17 Connell, R. S.; Swank, R. L.; Webb, M. C.: The development of pulmonary ultra-structural lesions during hemorrhagic shock. J. Trauma *15:* 116–129 (1975).

18 Ratliff, N. B.; Wilson, J. W.; Mikat, E.; Hackel, D. B.; Graham, T. C.: The lung in hemorrhagic shock. IV. The role of the polymorphonuclear leukocyte. Am. J. Path. *65:* 325–334 (1971).

19 Teplitz, C.: The core pathobiolog and integrated medial science of adult acute respiratory insufficiency. Surg. Clins N. Am. *56:* 1091–1133 (1976).

20 Wilson, J. W.: Some effects of shock on the lung's cellular components; in The cell in shock. Proc. Recent Research Developments and Current Clinical Practice in Shock, pp. 39–44 (Upjohn, 1974).

21 Redl, H.; Schlag, G.; Hammerschmidt, D. E.: Quantitative assessment of leukostasis in experimental hypovolemic-traumatic shock. Acta chir. scand. *150:* 113–117 (1984).

22 Dinges, H. P.; Redl, H.; Schlag, G.: Quantitative assessment of granulocytes in the lung after polytrauma – dog and human autopsy data. Eur. surg. Res. *16:* 100–101 (1984).

23 Craddock, P. R.; Hammerschmidt, D. E.; White, J. G.; Dalmasso, A. P.; Jacob, H.S.: Complement (C5a)-induced granulocyte aggregation in vitro: a possible mechanism of complement-mediated leukostasis and leukopenia. J. clin. Invest. *60:* 260–264 (1977).

24 Hammerschmidt, D. E.; Weaver, L. J.; Hudson, L. D.; Craddock, P. R.; Jacob, H. S.: Association of complement activation and elevated plasma-C5a with adult respiratory distress syndrome: pathophysiologic relevance and possible prognostic value. Lancet *i:* 947–949 (1980).

25 Heideman, M.; Saravis, C.; Clowes, G. H. A.: Effect of nonviable tissue and abscesses on complement depletion and the development of bacteremia. J. Trauma *22:* 527–532 (1982).

26 Heideman, M.: Complement activation in vitro induced by endotoxin and injured tissue. J. surg. Res. *26:* 670–674 (1979).

27 Holden, W. D.; DePalma, R. G.; Drucker, W. R.; et al.: Ultrastructural changes in hemorrhagic shock. Electron microscopic study of liver, kidney and striated muscle cells in rat. Ann. Surg. *162:* 517–520 (1965).

28 DePalma, R. G.; Holden, W. S.; Robinson, A. V.: Fluid therapy in experimental hemorrhagic shock: ultrastructural effects in liver and muscle. Ann. Surg. *175:* 539–546 (1972).

29 Schlag, G.; Redl, H.: The morphology of the microvascular system in shock: lung, liver and skeletal muscles. Crit. Care Med. (in press).

30 Van der Kleij, A. J.; De Koning, J.; Beerthuizen, G.; Goris, J. A.; Kreuzer, F.: Early detection of hemorrhagic hypovolemia by muscle oxygen pressure assessment: preliminary report. Surgery, St Louis *93:* 519–524 (1983).

31 Crowell, J. W.; Jones, C. E.; Smith, E. E.: Effect of allopurinol on hemorrhagic shock. Am. J. Physiol. *216:* 744–748 (1969).

32 Haglund, U.; Lundgren, O.: Intestinal ischemia and shock factors. Fed. Proc. *37:* 2729–2733 (1978).

33 Redl, H.; Schlag, G.: Permeabilitätsuntersuchungen an einem hypovolämisch-traumatischen Schockmodell bei Spontanatmung unter Verwendung einer Ringer- beziehungsweise Albuminlösung als Volumssubstitution. Hft Unfallheilk. *156:* 572–583 (1983).

34 Carlson, E. L.; Selinger, S. L.; Utley, J.; Hoffman, J. I.: Intramyocardial distribution of blood flow in hemorrhagic shock in anaesthetized dogs. Am. J. Physiol. *230:* 41–49 (1976).

35 Jones, C. E.; Smith, E. E.; Dupont, E.; Williams, R. D.: Demonstration of nonperfused myocardium in late hemorrhagic shock. Circul. Shock *5:* 97–104 (1978).

36 Proctor, H. J.; Starek, P. J.; Fry, J.: An investigation of endocardial viability ratio in myocardial failure following prolonged hemorrhagic shock. Ann. Surg. *181:* 893–895 (1985).

Prof. Dr. Günther Schlag, Ludwig Boltzmann Institute of Experimental Traumatology, Donaueschingenstrasse 13, A-1200 Vienna (Austria)

Novelli, Ursini (eds.), Oxygen Free Radicals in Shock. Int. Workshop,
Florence 1985, pp. 109–113 (Karger, Basel 1986)

Multiple Organ Failure and Sepsis without Bacteria: An Experimental Model

R. J. A. Goris[a], J. K. S. Nuytinck[a], W. K. F. Boekholtz[a],
I. P. T. v. Bebber[a], P. H. M. Schillings[b]

[a] Department of General Surgery, [b] Department of Pathologic Anatomy,
University Hospital St. Radboud, Nijmegen, The Netherlands

Failure of multiple organ systems (MOF) presently is the most common cause of death in severely injured patients and peritonitis patients admitted in an ICU. Since 'sepsis' has been first described by *Ashbaugh* et al. [1] as a highly lethal complication occurring in acute respiratory distress (ARDS) patients ventilated mechanically, no substantial improvement has been made in understanding or treating this syndrome. Bacteria and endotoxins have been almost invariably associated with this syndrome, though a consistent correlation with positive blood cultures and bacterial foci was found in only 33% of trauma patients with MOF [2]. As 'sepsis' involves action of a bacterial stimulus as well as the reaction of the body to such a stimulus, it seemed worthwile to study the bodily reaction in the absence of bacteria or endotoxins.

Material and Methods

2.5 g zymosan (Sigma Chemicals, St Louis, Mo., USA) was suspended by high frequency vibration in 100 ml of liquid paraffin. The suspension was sterilized by incubation in a water bath at 100 °C during 80 min. Preliminary experiments demonstrated that a dosage of 100 mg zymosan per 100 g of body weight resulted in severe illness in all rats, in an acute mortality of 35% within 24 h, and an additional mortality of 15% in the next days.

Experiment A. 40 male Wistar rats (body weight 300–400 g) were adapted to handling and to all experimental measurements. 4 days before the experiment, micro-electrodes were implanted subcutaneously on the back for ECG recording. The day of the experiment, the rats were randomized in three groups. Group I (20 rats) i.p. 100 mg

Table I. Results of blood analysis at day 12

| | Experiment A | | | p-values | | Experiment B | | p-values |
| | I | II | III | | | I | II | |
	zymosan	paraffin	blanco	I–II	II–III	zymosan	paraffin	I–II
Alkaline phosphatase, U/l	83±6	63±2	64±2	***		78±5	63±3	**
SGPT, U/l	28±1	23±2	30±1	**	**	30±2	27±1	
Lactate, μmol/l	4,009±110	3,143±295	2,835±274	*		4,089±507	2,317±255	**
Thrombocyte count $\times 10^9$/l	1,523±83	904±70	819±39	***		1,710±61	895±21	***
Leucocyte count $\times 10^9$/l	11.4±1.8	13.7±1.0	8.4±0.6		**	11.4±1.1	7.4±0.5	**
PMN, %	57±4	21±3	15±3	***		55±2	27±4	***

* $p \leq 0.05$; ** $p \leq 0.01$; *** $p \leq 0.001$ (Wilcoxon's test).

zymosan/100 g body weight in 4 ml paraffin. Group II (10 rats) i.p. 4 ml of paraffin. Group III (10 rats) blanco controls.

In all rats, the following measurements were made on day 0, 2, 6, 9 and 12. Oxygen consumption (VO_2) was measured in a volumetric respirator. Respiratory rate was determined visually, heart rate electrocardiographically and rectal temperature digitally.

At day 12 all surviving rats were bled by heart puncture and placed in a closed space saturated with ether. Blood was utilized for determination of thrombocytes, leucocyte count, differential count, alkaline phosphatase, serum glutamic pyruvic transaminase (SGPT), creatinine and lactate.

The thus sacrificed rats were weighed. Peritoneal fluid was collected for bacterial culture. The lungs with trachea, kidneys, liver and spleen were dissected free, weighed, and fixated with formaldehyde 4%.

Experiment B. 20 germ-free male Wistar rats (weight 300 g) were kept under germ-free conditions. All experimental procedures were identical to experiment A, except for the implantation of ECG electrodes and measurement of heart rate, respiratory rate, body temperature and VO_2. At day 0, rats were randomized in two groups. Group I received i.p. 100 mg zymosan/100 g body weight, group II 4 ml paraffin i.p.

Experiment C. 12 Wistar rats (body weight 200–300 g) were randomly given i.p. zymosan (as in experiment A group I) or served as blanco controls. After 14 days the rats were sacrificed by heart puncture and determinations of malon dialdehyde (thiobarbiturate method) performed on plasma, liver tissue and lung tissue.

Statistical analysis was performed with Wilcoxon's two sample test, Friedman's test, and Kruskal-Wallis' test. Results between groups were significant if $p < 0.05$.

Results

All group I rats of the experiments A, B and C showed symptoms of severe illness. They became lethargic, anorectic, hyperventilated, lost haemorrhagic fluid from the nose and conjunctivae, and lost liquid stools. In this phase 7 group A rats and 1 germ-free rat (group B) died. After 3 days, the rats improved. They became more active, and lost no more haemorrhagic fluid or fluid stools. Also respiration was closer to normal. After 5 days, the rats deteriorated progressively with increasing hyperventilation and dyspnoea, elevated heart rates and decreasing oxygen consumption.

In experiment A, 7 group I rats died within 20–72 h after inoculation, and 3 rats died after 9 – 12 days. In experiment B, 1 group I rat died early. In all 7 experiment A group I rats that died early and in 1 group I rat of experiment A terminated at day 12, positive bacterial cultures were obtained from the peritoneal fluid. Bacterial cultures of the peritoneal fluid remained negative in all other rats of both experiments. The results of blood analysis in the 12-day survivors are summarized in table I. Or-

Table II. Results of malondialdehyde (MDA) determination in plasma and tissue[1]

	Zymosan (n=6)	Blanco control (n=6)
MDA in plasma	±18.8 nmol/ml	±3.4 nmol/ml
MDA in liver tissue	±105.0 nmol/g wet weight	±5.4 nmol/g wet weight
MDA in lung tissue	±199.0 nmol/g wet weight	±14.8 nmol/g wet weight

[1] These experiments and determinations were performed in the Ludwig Boltzmann Institute, Vienna (Prof. Dr. *G. Schlag* and Dr. *H. Redl*).

gan weights of lungs, liver, kidneys, and spleen all were significantly increased, as compared to group II in both experiment A and B. The group II rats of experiment A also had significantly, though only slightly, more elevated lung weights than blanco controls.

In the group I rats of experiments A and B, microscopically, the lungs showed interstitial and intra-alveolar oedema, strongly increased numbers of intracapillary granulocytes and an increase in foam cells, probably macrophages. The liver showed congestion and oedema, an increase in the number of granulocytes inside the capillaries and in Disse's space. The spleen showed an increase in granulocytes in the capillaries and in the pulpa, and exhaustion of lymphoid tissue. In the kidneys widening of tubuli was seen, with granulation of cytoplasm. The results of experiment C are shown in table II.

Discussion

Zymosan is a potent activator of the alternate pathway of the complement system. As zymosan is not water-soluble, the experimental model concerns local stimulation of inflammation. In these experiments, intraperitoneal zymosan resulted in a triphasic illness, resembling MOF.

The changes found in this non-bacterial model strikingly resemble the findings in rat models of 'sepsis' [3–6]. *Wichterman* et al. [3] described the caecal ligation and puncture method in rats, resulting in an early hyperdynamic phase with fever, lethargy, tachypnoea, increased heart rate, and in a late hypodynamic phase with high serum lactate levels. The non-bacterial model presented in this article also conforms to the 'guidelines for progressive and lethal sepsis models' proposed by *Wichterman* et al. [3] except for the presence of micro-organisms. In experiment B, it is demonstrated that an ARDS and MOF-like syndrome with 'sepsis' can be induced in germ-free rats by a non-bacterial, non-endotoxic stimulus.

Important also is the finding that in 8 experiment A group I rats, positive bacterial cultures were obtained from the peritoneal cavity, 7 even within a few days after inoculation. Somehow, massive stimulation of inflammation within the peritoneal cavity thus results in bacterial colonization of the peritoneal cavity. The subsequent increased inflammatory response to the bacterial stimulus might have led to bacterial sepsis and early death in these rats.

The highly significant increases in malon dialdehyde content of plasma, liver and lung tissue, demonstrate that in this experimental model a high degree of lipid peroxydation is present.

This study provides evidence for our hypothesis that MOF is the result of a generalized autodestructive process of inflammatory character [2].

References

1   Ashbaugh, D. G.; Petty, T. L.: Sepsis complicating the acute respiratory distress syndrome. Surgery Gynec. Obstet. *135:* 865–868 (1972).
2   Goris, R. J. A.; teBoekhorst, T.; Nuytinck, J. K. S.: Multiple organ failure. Generalized autodestructive inflammation? Archs Surg., Chicago *120:* 1109–1115 (1985).
3   Wichterman, K. A.; Baue, A. E.; Chaudry, I. H.: Sepsis and septic shock. A review of laboratory models and a proposal. J. surg. Res. *29:* 189–201 (1980).
4   Lang, C. H.; Bagby, G. J.; Bornside, G. H.; Vial, L. J.; Spitzer, J. J.: Sustained hypermetabolic sepsis in rats: characterization of the model. J. surg. Res. *35:* 201–210 (1983).
5   Oh, G. R.; Mela-Riker, L. M.; Bryant, R. E.; Lowe, D. K.: A new experimental model of chronic high-output sepsis. Circul. Shock *13:* 99 (1984).
6   Kirton, O. C.; Jones, R.; Zapol, W. M.; Reid, L.: The development of a model of subacute lung injury after intra-abdominal infection. Surgery, St. Louis *96:* 384–394 (1984).

Prof. R. J. A. Goris, Department of General Surgery, University Hospital St. Radboud, PO Box 9101, NL-6500 HB Nijmegen (The Netherlands)

Novelli, Ursini (eds.), Oxygen Free Radicals in Shock. Int. Workshop,
Florence 1985, pp. 114–118 (Karger, Basel 1986)

The Effect of Oxygen Free Radical Scavengers on Experimental Endotoxemia

F. Kunimoto, T. Morita, R. Ogawa

Red Cross Medical Center, Tokyo, Japan

It has been reported that excessive lipid peroxidation played a key role in the pathogenesis of lethality in experimental endotoxemia [1, 2]. The inhibitory measures of excessive lipid peroxidation is expected to improve the survival rate of animals subjected to endotoxemia.

In the present study, three measures were tried to protect from excessive lipid peroxidation in endotoxemic rats. The first measure was the pretreatment of xanthine oxidase (XOD) inhibitor, allopurinol, to suppress the production of superoxide anion (O_2^-). The second measure was the administration of enzymatic quencher of oxygen radicals such as superoxide dismutase (SOD) and/or catalase (CAT). The third measure was the supplementation of exogenous chemical scavengers such as α-tocopherol (VE) or reduced glutathione (GSH).

Materials and Methods

The experimental endotoxemia was made in Wistar strain rats weighing 250 g by an intraperitoneal injection of a lethal dose of *E. coli* endotoxin (0.4 mg/100 g body weight). The endotoxin was extracted from *E. coli* 055:18.

The study consisted of three experiments. In the first experiment, the maximal salutary doses of three measures were determined. Allopurinol dissolved in polyethylene glycol was given intraperitoneally in doses of 1, 5, 10 mg/100 g 30 min prior the injection of endotoxin as a first measure. SOD in doses of 300, 900, 9,000 U/100 g and/or CAT of 40, 400, 4,000 U/100 g were administered intra-peritoneally 30 min prior to endotoxin. VE in doses of 4, 10, 100 mg/100 g or GSH of 5, 50, 100 mg/100 g were pretreated 30 min prior to endotoxin as third measures. The number of rats alive in each group as recorded every 3 h until 24 h after injection of endotoxin. In the second experiment, lipoperoxide concentration, XOD activity [3] and SOD activity [4] in the liver of endotoxemic rats which received maximal effective doses of three measures 3 h after injection of endotoxin.

In the third experiment, intracellular distribution of lysosomal hydrolases [5, 6] were determined in livers of rats which were pretreated with maximal doses of three measures after 3 h of endotoxin injection. The analysis was composed of homogenization and centrifugation of liver in buffered 0.25-M sucrose, and the ratio of soluble hydrolases in cytoplasm and bound hydrolases in pellets were adopted as an index.

Results

The effects of three measures on survival rate are shown in table I. The maximal salutary doses were 1 mg/100 g of allopurinol in the first measure, a combination of 9,000 U/100 g of SOD and 4,000 U/100 g of CAT in the second measure and 10 mg/100 g of VE in the third measure. A combination of SOD and CAT depicted the highest survival rate of the three measures.

Lipoperoxide concentration, XOD activity and SOD activity in the livers of rats received maximal effective doses of the three measures as shown in table II. The smaller dose of allopurinol, 1 mg/100 g, came forth highly inhibited XOD activity with lower hepatic lipoperoxide compared with controls. The combination of SOD and CAT produced lower concentrations of hepatic lipoperoxide concentration and higher SOD activity. Large amount of exogenous VE induced mild suppression of hepatic lipoperoxide concentration with no changes in XOD and SOD activity.

As shown in table III, the ratio of soluble and bound lysosomal hydrolases was maintained in the second and third measures of maximal salutary doses, showing the protective effect of measures from disruption of lysosomes.

Discussion

Lipid peroxidation is initiated by oxygen free radicals on biomembrane, propagating to the injury of biomembranes. In the hypoxic condition such as shock, ground state oxygen was excessively activated to free radicals by addition of electrons leaked out from several enzymatic systems. The xanthine-XOD system is known to be a principal system producing radicals in the cytoplasm. *Morita* et al. reported that XOD is acti-

Table I. Survival rates of the rats at 6h after endotoxin injection

A. *Control, allopurinol*

	Control	Allopurinol 1 mg/100 g	Allopurinol 5 mg/100 g	Allopurinol 10 mg/100 g
Survivals	8/30	7/20	7/20	5/20

B. *SOD, CAT*

	SOD 300 U/100 g	SOD 900 U/100 g	SOD 9,000 U/100 g	CAT 40 U/100 g	CAT 400 U/100 g	CAT 4,000 U/100 g	SOD 9,000 U/100 g + CAT 4,000 U/100 g
Survivals	6/20	7/20	14/20	5/20	5/20	6/20	15**/20

C. *VE, GSH*

	VE 4 mg/100 g	VE 10 mg/100 g	VE 100 mg/100 g	GSH 5 mg/100 g	GSH 50 mg/100 g	GSH 100 mg/100 g
Survivals	4/20	7/20	9/20	5/20	8/20	6/20

Abbreviations defined in the text.
** p < 0.01.

Table II. Lipoperoxide concentrations, SOD and XOD activities at 3 h after endotoxin injection (M±SE)

Pretreatment	Lipoperoxide concentration nM/g	SOD activity U/g	XOD activity mU/g
Saline (n = 10)	47.5 ± 1.0	8,204 ± 158	283.6 ± 18.6
Control (n = 12)	94.5 ± 11.3	5,470 ± 336	907.2 ± 74.8
Allopurinol, 1 mg/100 g (n = 12)	72.4 ± 8.6**	6,630 ± 467**	103 ± 7.9**
SOD, 9,000 U/100 g (n = 18)	55.3 ± 4.6**	9,700 ± 645**	320 ± 22.3**
CAT, 4,000 U/100 g (n = 10)	103 ± 8.8	5,560 ± 499	942 ± 124
SOD, 9,000 U/100 g + CAT 4,000 U/100 g (n = 15)	57.6 ± 5.2**	10,400 ± 472**	376 ± 22.7**
VE, 100 mg/100 g (n = 15)	62.3 ± 5.2**	5,230 ± 453	857 ± 72.3
GSH, 50 mg/100 g (n = 14)	69.5 ± 6.1**	5,720 ± 487	905 ± 83.6

** $p < 0.01$ compared with control group.

Table III. The ratio of soluble and bound lysosomal hydrolases in the liver cell (soluble/bound; M±SE)

Pretreatment	β-Glucuronidase	Cathepsin D
Saline (n = 10)	0.65 ± 0.07	0.56 ± 0.04
Control (n = 8)	1.41 ± 0.18	0.88 ± 0.08
Allopurinol, 1 mg/100 g (n = 12)	1.30 ± 0.11	0.80 ± 0.07
SOD, 9,000 U/100 g (n = 14)	0.69 ± 0.04**	0.53 ± 0.03**
VE, 100 mg/100 g (n = 10)	0.82 ± 0.09**	0.68 ± 0.05**
GSH, 50 mg/100 g (n = 12)	0.95 ± 0.08**	0.71 ± 0.05**

** $p < 0.01$ compared with control group.

vated in the liver of rats with endotoxemia. Inhibition of XOD is a pro-
totypical measure to protect tissues from lipid peroxidation as shown in
the present study. However, a smaller dose of allopurinol was more effective than larger doses, suggesting an essential role of oxygen free radicals
as the substrate of biosynthesis of many active substances.

In the mammalian cells, oxygen free radicals were maintained at low
concentrations by the eliminating action of SOD. The exogenous administration of SOD has been expected to be an effective measure as a dismutating agent of superoxide anion. In the present study, the combination of SOD and CAT produced a maximal salutary effect of the three
measures. However, the precise mechanism remains to be solved, because exogenous SOD is belived not to penetrate cellular membrane.

VE and GSH are chemical scavengers physiologically present in the
mammalian cell. Exogenous administration in large amounts showed salutary effects on experimental endotoxemia in the present study. Those
hydrophilic scavengers can easily penetrate the cellular membrane and
absorb the free radicals.

References

1   Ogawa, R.; Fujita, T.: Changes in hepatic lipoperoxide concentration in endotox-
    emic rats. Circulatory Shock 9: 369–374 (1982).
2   Tyler, D. D.: Role of superoxide radicals in the lipid peroxidation of intracellular
    membrane. FEBS Lett. 51: 180–183 (1975).
3   Hashimoto, S.: A new spectrophotometric assay method of xanthine oxidase in
    crude tissue homogenate. Annls Biophys. 62: 426–435 (1974).
4   Misra, H.; Fridovich, I.: The role of superoxide anion in the autoxidation of
    epinephrine and simple assay for superoxide dismutase. J. biol. Chem. 247: 3170
    (1982).
5   Fishman, W.; Springer, B.; Branetti, L.: Application of an improved glucuronidase
    assay, method to the study of human blood β-glucuronidase. J. biol. Chem. 173:
    449–456 (1948).
6   Anson, M.: Estimation of cathepsin, hemoglobin and partial purification of
    cathepsin. J. gen. Physiol. 20: 565 (1936).

F. Kunimoto, MD, Red Cross Medical Center, 4-1-22 Hiroo Shipuya-ku,
Tokyo 371 (Japan)

Novelli, Ursini (eds.), Oxygen Free Radicals in Shock. Int. Workshop,
Florence 1985, pp. 119–124 (Karger, Basel 1986)

Anti-Shock Action of Phenyl-*t*-Butyl-Nitrone, a Spin Trapper

*Gian Paolo Novelli, Patrizia Angiolini, Guglielmo Consales,
Roberto Lippi, Riccardo Tani*

Institute of Anaesthesiology and Intensive Care, University of Florence, Italy

Introduction

A conclusive support of the role of oxygen free radicals in the path-ogenesis of shock might derive from experiments in conditions of absence of oxygen free radical generation. However, this condition is impossible to be obtained in living animals. The capture of free radicals immediately after generation and before impact on biological structures might be very similar to that without interference in cell respiration.

This capture ('trapping') is performed by some chemical compounds named 'spin trappers', that interact with free radicals forming a more stable adduct. The electron spin resonance of the adduct is used to detect and measure free radicals with an half-life too brief to be detected [*Rosen* et al., 1984].

The spin trapper used in present experiments was phenyl-*t*-butyl-ni-trone (PBN) that forms a nitroxide adduct with O_2^- and $OH^.$.

In previous experiments, a daily administration of PBN for 6 days appeared to be deprived of gross behavioral or anatomic effects [*Novelli* et al., 1984].

The present experiments were directed to search for: (1) the effects on the survival of PBN injected *before* a 100% lethal dose of endotoxin, trauma or superior mesenteric artery occlusion (SMAO); (2) the effects on the survival, acid-base status and hematocrit of PBN injected 30 or 60 min *after* a 100% lethal dose of trauma.

Methods

Experiments were performed on male Wistar rats (200 g) fasted 12 h before the experiments. Endotoxin shock was provoked by 20 mg·kg^{-1} i.p. of *Escherichia coli* lipopolysaccharide (Difco W 0111: B_4). Whole body trauma shock was provoked by a rotating drum (58 revolutions·min^{-1}; 1,200 revolutions). The rats dead in the drum were discarded from the experiments, as their death was supposed to be due to an acute event different from shock. SMAO shock was provoked by arterial occlusion (50 min) under ketamine anesthesia. At the end of the shock-provoking maneuvers, animals were allowed to drink water ad libitum.

PBN (Kodak, 11363) was suspended in olive oil immediately before administration and minimizing exposition to light and air; suspension was prepared so as to have each single dose in 1 ml. The control rats were given only olive oil. Acid-base status was evaluated on arterial blood with a micromethod (IL 1312). Hematocrit value was measured in duplicate with a micromethod. Statistical evaluations were performed with the χ^2 test, comparing each result with the control one, at each experimental time. Acid-base and hematocrit values were reported as mean values ± SE.

Experiments

(1) PBN 150-100-50 mg·kg^{-1} was injected i.p. to rats 10 min before trauma or endotoxin; in SMAO experiments PBN was administered 20 min before laparatomy. Survival was measured at 3, 6, 12, 24 h after the end of the shock-provoking maneuvers.

(2) Rats submitted to whole body trauma. After 30 or 60 min PBN (150 mg·kg^{-1}) or olive oil was injected i.p. and survival was controlled. Acid-base status and hematocrit were examined on rats: (a) untreated; (b) at the end of trauma; (c) 3 h after b; (d) 6 h after b. Blood was taken by cardiac puncture and these animals were sacrificed.

Results

Survival of rats submitted to endotoxin after PBN was enormously higher than that of non-pretreated rats. The lowest dose afforded incomplete protection (table I). In trauma experiments all PBN-pretreated rats survived (table II). In SMAO experiments also, PBN was protective against shock (table III).

The second series of experiments was directed to search if the spin trapper was effective in reversing traumatic shock, whose severity was confirmed by metabolic acidosis and high hematocrit value (tables IV/V). Injection of PBN until 60 min after trauma was effective in preventing death (table IV) and ameliorating both acid-base and hematocrit.

Table I. Survival after injection of *E. coli* endotoxin (Difco W 0111:B_4, 20 mg·kg^{-1} i.p.) in rats pretreated with olive oil i.p. or with the spin trapper phenyl-*t*-butyl-nitrone (PBN) suspended in oil

		Hours after endotoxin			
		3	6	12	24
Olive oil	survivors	10/15	4/15	1/15	1/15
(n = 15)					
PBN 150 mg·kg^{-1}	survivors	15/15	15/15	15/15	15/15
(n = 15)	p <	0.01	0.00001	0.00001	0.00001
PBN 100 mg·kg^{-1}	survivors	14/15	14/15	14/15	14/15
(n = 15)	p <	0.06	0.0001	0.00001	0.00001
PBN 50 mg·kg^{-1}	survivors	11/15	11/15	10/15	10/15
(n = 15)	p <	n.s.	0.01	0.001	0.0005

Statistical evaluation was performed with the χ^2 test, comparing the survival of PBN-pretreated rats with that of control ones at each experimental time.

Table II. Survival after whole body trauma (rotating drum; 1,200 revolutions) in rats pretreated with olive oil i.p. or with the spin trapper phenyl-*t*-butyl-nitrone (PBN) suspended in oil

		Hours after the end of the rotation			
		3	6	12	24
Olive oil	survivors	8/13	5/13	1/13	0/13
(n = 13)					
PBN 150 mg·kg^{-1}	survivors	14/14	14/14	14/14	14/14
(n = 14)	p <	0.01	0.0001	0.00001	0.000001
PBN 100 mg·kg^{-1}	survivors	14/14	14/14	14/14	14/14
(n = 14)	p <	0.01	0.0001	0.00001	0.000001
PBN 50 mg·kg^{-1}	survivors	10/12	10/12	9/12	9/12
(n = 12)	p <	n.s.	0.02	0.001	0.0001

Rats were introduced into the rotating drum as groups of 5 and those which died during rotation were discarded from calculations; therefore, experimental groups were numerically inhomogenous.

Table III. Survival after superior mesenteric artery occlusion (SMAO; 50 min occlusion) in rats pretreated with olive oil i.p. or with spin trapper phenyl-*t*-butyl-nitrone (PBN) suspended in oil

		Hours after SMAO			
		3	6	12	24
Olive oil (n = 15)	survivors	10/15	8/15	0/15	0/15
PBN 150 mg·kg⁻¹ (n = 15)	survivors	15/15	15/15	15/15	15/15
	$p <$	0.01	0.0001	0.000001	0.000001
PBN 100 mg·kg⁻¹ (n = 15)	survivors	15/15	14/15	10/15	10/15
	$p <$	0.01	0.001	0.0001	0.0001
PBN 50 mg·kg⁻¹ (n = 15)	survivors	13/15	12/15	9/15	7/15
	$p <$	n.s.	n.s.	0.001	0.01

Statistical evaluation was performed with the χ^2 test, comparing the survival of PBN pretreated rats with that of control ones at each experimental time.

Table IV. Survival in rats treated with olive oil i.p. or with the spin trapper phenyl-*t*-butyl-nitrone (PBN 150 mg·kg⁻¹) in oil, after 10, 30 or 60 min to the end of whole body trauma (rotating drum; 1,200 revolutions)

		Hours after the end of the rotation			
		3	6	12	24
Olive oil (n = 16)	survivors	8/16	0/16	0/16	0/16
PBN after 30 min (n = 14)	survivors	14/14	14/14	14/14	14/14
	$p <$	0.001	0.000001	0.000001	0.000001
PBN after 60 min (n = 16)	survivors	16/16	16/16	16/16	16/16
	$p <$	0.001	0.000001	0.000001	0.000001

Discussion

The biological actions of PBN are completely unknown: in fact this compound is used mostly in in vitro experiments and only rarely has it been used in vivo as a single dose to detect certain kinds of free radicals. Our experiments also failed to show any gross effects after repeated administration of PBN to rats.

Table V. Acid-base status of rats submitted to 100% lethal traumatic shock given PBN 150 mg·kg⁻¹ i.p. olive oil

	3 h after trauma	6 h after trauma
Olive oil (n = 10)	pH 7.08 ±0.05	all died
	BE −12.8 ±0.2	all died
	Hct 58.7 ±1.4	all died
PBN injected 30 min after trauma (n = 5)	pH 7.20 ±0.04	7.30±0.1
	BE −12.0 ±0.5	−5.8±0.2
	Hct 52.2 ±1.3	48.1±1.8
PBN injected 60 min after trauma (n = 5)	pH 7.20 ±0.08	7.28±0.06
	BE −12.2 ±0.4	−9.1±0.8
	Hct 56.1 ±1.6	53.0±2.1
	normal values	immediately after trauma
	pH 7.40 ±0.2	7.18±0.07
	BE 0 ±2	−19±1.7
	Hct 47% ±0.9	56±1.4

Control experiments were performed injecting rats with olive oil alone at the same experimental times as PBN. Results were not different and therefore were cumulated.

Although it is largely improbable that trapping of free radicals is deprived of biological effects, it must be accepted that in our knowledge the only known action of PBN is that of trapping oxygen free radicals. As a consequence, the reported results are to be interpreted as proofs of the role of oxygen free radicals in the pathogenesis of shock; in fact, their capture and temporary inactivation completely prevents three types of otherwise lethal shock.

It is remarkable that PBN administered to shocked rats was also effective in preventing death from traumatic shock, whose severity was documented by the metabolic acidosis (pH 7.18±0.07; BE −19±1.17) and by the hematocrit value (56±0.4).

All the untreated rats died in a very brief time but all those injected with the spin trapper after 60 min were permanent survivors; their acid-base status and hemoconcentration returned toward normal values without any therapy.

References

Novelli, G. P.; Bordi, L.; De Gaudio, A. R.: Trapping of free-radicals prevents endo-toxin shock. 1st Meet. European Shock Society, Manchester 1984.

Rosen, G. M.; Rauckman, E. J.: Spin trapping of superoxide and hydroxyl radicals. Meth. Enzym. *105:* 198–209 (1984).

Prof. G. P. Novelli, Institute of Anaesthesiology and Intensive Care, University of Florence, Policlinico di Careggi, Viale Morgagni, I-50134 Florence (Italy)

Novelli, Ursini (eds.), Oxygen Free Radicals in Shock. Int. Workshop,
Florence 1985, pp. 125–136 (Karger, Basel 1986)

Endotoxin-Induced Disseminated Intravascular Coagulation and Shock, and Their Relationship with Oxygen-Derived Free Radicals

T. Yoshikawa, O. Seto, K. Itani, Y. Kakimi, S. Sugino, M. Kondo

First Department of Medicine, Kyoto Prefectural University of Medicine,
Kamikyo-ku, Kyoto, Japan

It is now widely recognized that oxygen free radicals are implicated in a wide variety of in vivo biological reactions. Complete reduction of a molecule of oxygen to water requires four electrons, and in sequential univalent process several intermediates will be encountered. These are the superoxide anion radical, hydrogen peroxide, and the hydroxyl radical [1]. They are highly reactive molecules which could potentially damage surrounding cellular structures and subcellular organelles. The primary defence is provided by enzymes that catalytically scavenge the intermediates of oxygen reduction. The superoxide radical is eliminated by superoxide dismutase (SOD), which catalyzes its conversion to hydrogen peroxide plus oxygen [2], and hydrogen peroxide is removed by catalase [3], which converts it to water plus oxygen. These protective mechanisms have evolved to defend cell components from free radical damage, but disease states, xenobiotics, and other environmental stresses can overwhelm defence mechanisms and cause cytotoxicity. The protective activity of enzymatic scavengers is diminished by acidosis [4] and by endotoxin [5]. It was reported that an exposure to endotoxin can prime neutrophils for increased release of superoxide [6], and that an endotoxin challenge results in superoxide generation in mice at levels that may ultimately result in death [7].

A generation of oxygen free radicals seems to be implicated in the pathogenesis of shock and disseminated intravascular coagulation (DIC) caused by endotoxin. To elucidate the hypothesis, the effects of SOD and catalase on endotoxin-induced shock and DIC were examined in rats.

Materials and Methods

Female rats of Wistar strain with weights ranging from 200 to 220 g were housed for at least 7 days in our animal quarters prior to the experiments. They were kept at a constant ambient temparature of 20 °C and were fed a standard diet (Oriental Yeast Co., Tokyo, Japan) and water ad libitum. Endotoxin (*Escherichia coli* 055:B5 lipopolysaccharide B; Difco Lab., Detroit, Mich., USA) was dissolved in pyrogen-free physiological saline before every experiment. Experimental DIC was induced by sustained infusion of 100 mg/kg of endotoxin, diluted in 11.4 ml of saline, into the femoral vein for 4 h using a syringe. Control animals were infused with 11.4 ml of pyrogen-free saline in a manner identical with that used for animals receiving endotoxin. For i.v. infusions the femoral veins of the animals were cannulated unter light Nembutal (5 mg/100 g, i.p.) anesthesia. Experimental shock was induced by a single injection of 100 mg/kg of endotoxin, diluted in 1.0 ml of physiological saline, into the femoral vein using a syringe.

To examine the effects of SOD and catalase on the experimental DIC, the rats were injected with SOD from bovine liver (3,500 U/mg Orgotein; Toyo Jozo Co., Ltd. Sizuoka, Japan) at 0.5, 5.0 or 50.0 mg/kg or catalase from bovine liver (Hydrogen peroxide oxidoreductase, EC 1.11.1.6, 35,000 Sigma U/mg protein; Sigma Chemical Co., St. Louis, Mo., USA) at 0.01, 0.1 or 1.0 mg/kg, subcutaneously; immediately after the injection of these agents, the animals were infused continuously with 100 mg/kg of endotoxin for 4 h. To examine the effect of SOD and catalase on experimental shock, the rats were injected with SOD from bovine blood (3,050 U/mg protein; Sigma Chemical Co., St. Louis, Mo., USA) at 10.0 or 50.0 mg/kg or catalase from bovine liver (40,000 Sigma U/mg protein; Sigma Chemical Co., St. Louis, Mo., USA) at 0.1 or 1.0 mg/kg, subcutaneously 12 and 1 h before the injection of endotoxin (100 mg/kg).

The severity of DIC was determined with various parameters, such as fibrinogen and fibrin degradation products (FDP), fibrinogen, prothrombin time (PT), partial thromboplastin time (PTT), platelet count, and percent glomerular fibrin deposits (% GFD); 100 glomeruli were counted and those having fibrin thrombi were expressed in percentage. The severity of shock was determined with systolic blood pressure, heart rate, and lysosomal enzymes, such as acid phosphatase, β-glucuronidase and cathepsin B.

FDP was determined by the latex agglutination test (Teikoku Hormone Mfg. Co., Tokyo, Japan) and reported in µg/ml of equivalent to human fibrinogen base [8]. Fibrinogen was determined by the technique of *Ratnoff and Menzie* [9]. PT and PTT were measured according to the methods of *Quick* et al. [10], and *Langdell* et al. [11]. Platelets were counted by phase-contrast microscopy [12]. Kidney tissue specimens were examined histologically by hematoxylin-eosin and Wigert's fibrin stain [13]. Systolic blood pressure and heart rate count were measured by a tail-cuff method [14] using a sphygmomanometer PS-100 (Riken Kaihatsu Co., Tokyo, Japan). Serum acid phosphatase activity was assayed according to the method of *Andersch and Szczypinski* [15] using *p*-nitrophenylphosphate as the substrate. Serum β-glucuronidase activity was measured with Sigma reagents (Sigma Chemical Co., St. Louis, Mo., USA), based essentially on the method of *Fishman* et al. [16] using phenolphthalein glucuronic acid as the substrate. Cathepsin B was assayed by the method of *Otto and Bhakdi* [17] using α-*N*-benzoyl-*DL*-arginine-4-nitro anilide as a substrate.

Thiobarbituric acid (TBA) reactive substances in serum were determined by the method of *Yagi* [17]. The serum lipids were precipitated with protein using phosphotungstic acid and made to react with thiobarbituric acid. The reaction products were measured fluorometrically with excitation at 515 nm and emission at 553 nm. The concentration of TBA-reactive substances was expressed in terms of malondialdehyde (nmol/ml) using tetramethoxypropane as a standard.

Results

Endotoxin-Induced DIC

At 1 or 2 h after the infusion of endotoxin, no marked changes were seen in any of the parameters, but at 3 h FDP and %GFD increased (table I), PT and PTT were prolonged, and fibrinogen and platelet count were reduced. At 4 h coagulation was further advanced and glomerular fibrin deposits were present in all rats. Serum TBA reactants were increased significantly in rats infused by endotoxin.

To examine the protective effect of SOD or catalase, all rats were sacrificed at 4 h after the infusion of endotoxin and examined the affected changes of several following parameters.

The Changes in FDP. The FDP level increased to 56.1 ± 12.1 µg/ml (mean $\pm$ SD) in rats given 100 mg/kg/4 h of endotoxin. When 0.5, 5.0 or 50.0 mg/kg of SOD, or 0.01, 0.1 or 1.0 mg/kg of catalase was administered once, before the infusion of endotoxin, the FDP level was reduced significantly (tables II, III).

The Changes in Fibrinogen. The levels of fibrinogen decreased to 0.2 g/1 or less after the infusion of 100 mg/kg/4 h of endotoxin. The decrease in fibrinogen level was inhibited significantly in rats administered 50.0 mg/kg of SOD or 1.0 mg/kg of catalase.

The Changes in PT. PT was prolonged to 32.4 ± 6.4 s (mean $\pm$ SD) after the 4 h infusion of endotoxin (100 mg/kg). The prolongation of PT was inhibited significantly in rats administered 50.0 mg/kg of SOD, or 0.1 or 1.0 mg/kg of catalase before the infusion of endotoxin.

The Changes in PTT. PTT was prolonged to 152.1 ± 17.5 s after the 4-hour infusion of endotoxin (100 mg/kg). Pretreatment with SOD at 0.5, 5.0 or 50.0 mg/kg, or catalase at 0.1 or 1.0 mg/kg inhibited the prolongation significantly.

The Changes in Platelet Counts. The platelet counts were decreased to $66 \pm 9 \times 10^9/1$ (mean $\pm$ SD) in the rats infused with endotoxin (100 mg/kg/4 h). When rats were administered 0.5, 5.0 or 50.0 mg/kg of

Table I. The changes in coagulation test values and serum TBA-reactive substances during 4 h sustained infusion of endotoxin (100 mg/kg/4 h)

Hour	FDP μg/ml	Fibrinogen g/l	PT s	PTT s	Platelet ×10^9/l	GFD[1] %	TBA reactants[2] nmol/ml	N[3]
0 (normal[4])	2.6±0.5	1.10±0.31	11.8±0.7	59.7±9.9	1084±196	0	3.2±0.5	10
1	3.1±0.5	0.96±0.18	13.5±0.9	65.6±11.1	649±79[5]	9.8±3.0[5]	4.6±0.7[5]	12
2	4.8±0.5	0.57±0.09[5]	14.0±0.9	67.0±15.9	468±56[5]	13.4±5.5[5]	4.9±1.3[5]	12
3	38.9±9.1[5]	0.05±0.01[5]	23.5±2.1[5]	112.7±30.2[5]	217±32[5]	59.2±11.8[5]	5.1±0.8[5]	12
4	53.5±11.2[5]	0.02±0.01[5]	24.8±6.0[5]	134.7±22.9[5]	89±38[5]	80.3±14.5[5]	5.9±1.2[5]	24

Results are expressed as mean value ± SD.
[1] Glomerular fibrin deposits. [2] Measured as nmol of malondialdehyde. [3] Number of subjects. [4] Normal rats (0 h) were not infused endotoxin. [5] $p < 0.001$ for difference to controls by Student's t-test.

Table II. Effects of SOD on experimental DIC in rats. SOD was injected subcutaneously before the infusion of endotoxin (100 mg/kg/4 h)

	FDP μg/ml	Fibrinogen g/l	PT s	PTT s	Platelet count ×10^9/l	GFD[1] %	N[2]
Control[3]	56.1±12.1	< 0.2[4]	32.4±6.4	152.1±17.5	66±9	87.6±12.4	12
SOD[5]							
0.5 mg/kg	28.4±6.4[8]	< 0.2	30.9±8.9	130.6±18.8[7]	79±18[7]	62.7±9.7[8]	8
5.0	20.8±7.7[8]	< 0.2	28.9±9.3	93.7±11.5[8]	124±44[8]	58.6±7.2[8]	8
50.0	12.4±6.2[8]	0.58±0.16[8]	19.1±4.2[8]	74.0±12.0[8]	221±71[8]	28.6±5.4[8]	16
Normal rats[6]	3.8±0.8[8]	1.14±0.20[8]	12.7±0.4[8]	59.6±4.8[8]	603±81[8]	0[8]	12

Results are expressed as mean value ± SD.
[1] Glomerular fibrin deposits. [2] Number of animals. [3] Saline (1.0 ml) was injected once subcutaneously before the infusion of endotoxin (100 mg/kg/4 h). [4] Fibrinogen level was less than 0.2 g/l. [5] SOD was dissolved in 1.0 ml of saline. [6] Normal rats were infused with 11.4 ml/4 h of saline alone. [7] $p < 0.05$. [8] $p < 0.001$ for difference to the value of control rats by Student's t-test.

SOD, or 0.01, 0.1 or 1.0 mg/kg of catalase before the infusion of endoxin, the reduction of platelet count was inhibited significantly in these animals compared with that in the rats administered with saline and infused with endotoxin.

Formation of Fibrin Thrombi in the Renal Glomeruli. In the histological examination of kidney specimens collected from rats infused with en-

Table III. Effects of catalase on experimental DIC in rats. Catalase was injected subcutaneously before the infusion of endotoxin (100 mg/kg/4 h)

	FDP µg/ml	Fibrinogen g/l	PT s	PTT s	Platelet count ×10⁹/l	GFD[1] %	N[2]
Control[3]	56.1±12.1	< 0.2[4]	32.4±6.4	152.1±17.5	66±9	87.6±12.4	12
Catalase[5]							
0.01 mg/kg	29.6±10.9[8]	< 0.2	31.1±13.4	153.8±24.6	135±17[8]	63.8±12.2[8]	8
0.1	17.6±8.1[8]	< 0.2	25.0±5.4[7]	81.5±20.2[8]	166±28[8]	60.3±10.8[8]	8
1.0	13.9±6.1[8]	0.48±0.11[8]	19.2±5.7[8]	82.1±19.1[8]	227±35[8]	44.9±8.0[8]	19
Normal rats[6]	3.8±0.8[8]	1.14±0.20[8]	12.7±0.4[8]	59.6±4.8[8]	603±81[8]	0[8]	12

Results are expressed as mean value ± SD.
[1-4, 6-8] For footnotes see table II.
[5] Catalase was dissolved in 1.0 ml of saline.

dotoxin (100 mg/kg/4 h) or saline (11.4 ml/rat/4 h), 100 glomeruli were counted to detect those containing fibrin thrombi. In the normal rats infused with physiological saline solution for 4 h, the %GFD was 0%; that is, no fibrin thrombi were formed in any of the glomeruli. When the rats were infused with 100 mg/kg/4 h of endotoxin, %GFD was as high as 87.6 ± 12.4% (mean ± SD). However, when rats were administered 0.5, 5.0 or 50.0 mg/kg of SOD or 0.01, 0.1 or 1.0 mg/kg of catalase and infused with 100 mg/kg of endotoxin for 4 h the formation of fibrin thrombi was reduced significantly.

Direct Effects of SOD or Catalase on Various Parameters. Investigation was made to determine whether SOD or catalase directly affected the measured values of the various parameters. Rats were injected subcutaneously with 0.5, 5.0 or 50.0 mg/kg of SOD or 0.01, 0.1 or 1.0 mg/kg of catalase. At 1, 2, 3 and 4 h after the injection of SOD or catalase, FDP, fibrinogen, PT, PTT, platelet count and the number of renal glomeruli with fibrin thrombi were compared with the values obtained from rats receiving physiological saline as a control. There was no noticeable change.

Endotoxin-Induced Shock

Immediately after the injection of endotoxin, systolic blood pressure was reduced and heart rate was increased (fig. 1). Acid phosphatase and β-glucuronidase activities were increased (fig. 2). These changes were most remarkable at 45 min after the injection of endotoxin. All rats were

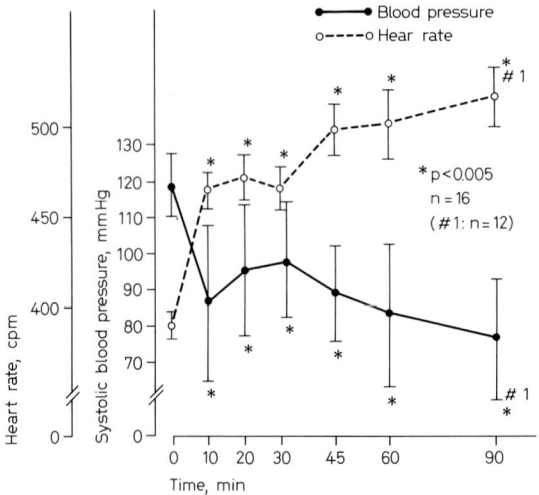

Fig. 1. Changes in systolic blood pressure and heart rate of rats injected with endotoxin (100 mg/kg).

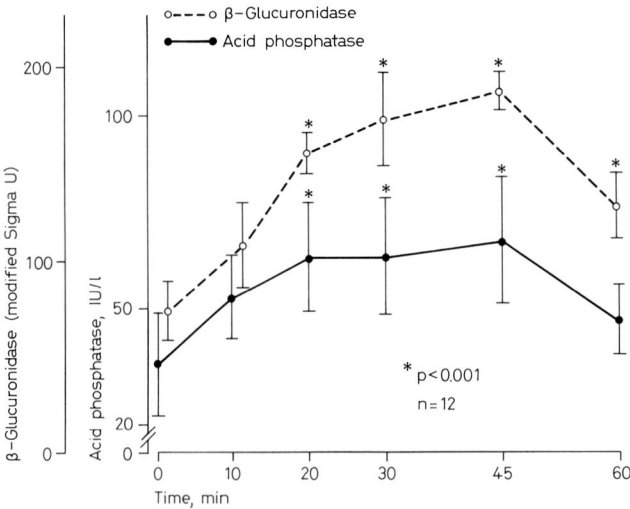

Fig. 2. Changes in serum acid phosphatase and β-glucuronidase activities after
the injection of endotoxin (100 mg/kg).

Table IV. Effects of SOD on experimental shock in rats. SOD was injected subcutaneously 12 and 1 h before the administration of endotoxin (100 mg/kg)

	Blood pressure mm HG	Heart rate cpm	Acid phosphatase IU/l	β-Glucuronidase U/ml	Cathepsin B OD/ml	TBA reactants nmol/ml	N[1]
Control[2]	71±15	490±35	68.5±11.7	181±71	3.1±0.5	3.6±0.3	16
SOD[3]							
10 mg/kg	101±15**	477±43	56.8±6.7**	201±59	2.9±0.8	3.3±0.4*	8
50	101±12**	486±45	48.4±6.8**	84±23**	2.1±0.5**	3.2±0.2**	8
Normal rats[4]	121±11**	404±29**	36.7±8.4**	58±18**	2.3±0.4**	2.9±0.3**	14

Results are expressed as mean value ± SD.
[1] Number of subjects.
[2] Physiological saline (1.0 ml) was injected s.c. 12 and 1 h before the injection of endotoxin (100 mg/kg).
[3] SOD was dissolved in physiological saline (1.0 ml).
[4] Normal rats were injected with 1.0 ml of physiological saline alone.
* $p < 0.01$; ** $p < 0.001$ for difference to the value of control rats by Student's t-test.

Table V. Effects of catalase on experimental shock in rats. Catalase was injected subcutaneously 12 and 1 h before the administration of endotoxin (100 mg/kg)

	Blood pressure mm Hg	Heart rate cpm	Acid phosphatase IU/l	β-Glucuronidase U/ml	Cathepsin B OD/ml	TBA reactants nmol/ml	N[1]
Control[2]	71±15	490±35	68.5±11.7	181±71	3.1±0.5	3.6±0.3	16
Catalase[3]							
0.1 mg/kg	79±21	445±55*	55.1±8.4*	150±64	2.8±0.7	3.4±0.5	8
1.0	105±18**	457±46*	45.1±6.7**	165±59	2.1±0.4**	3.5±0.3	12
Normal rats[4]	121±11**	404±29**	36.7±8.4**	58±18**	2.3±0.4**	2.9±0.3**	14

Results are expressed as mean value ± SD.
[1,2,4] For footnotes see table IV.
[3] Catalase was dissolved in physiological saline (1.0 ml).
* p < 0.01; ** p < 0.001 for difference to the value of control rats by Student's t-test.

killed at 45 min after the endotoxin injection, and examined the changes of following parameters of rats treated with SOD or catalase.

The Changes in Systolic Blood Pressure. Systolic blood pressure of rats injected with 1.0 ml of physiological saline alone was 121 ± 1 mm Hg (mean $\pm$ SD). The blood pressure was reduced significantly 45 min after the injection of endotoxin. The reduction was significantly inhibited by the 10 or 50 mg/kg of SOD, or 1.0 mg/kg of catalase (tables IV, V).

The Changes in Heart Rate. Heart rate was increased significantly by the injection of endotoxin. The increase was inhibited by the 0.1 or 1.0 mg/kg of catalase, but not by SOD.

The Changes in Serum Acid Phosphatase Activity. Serum acid phosphatase activity was increased 45 min after the injection of endotoxin. The increase in the activity was inhibited significantly by the administration of SOD (10 or 50 mg/kg) or catalase (0.1 or 1.0 mg/kg).

The Changes in Serum β-Glucuronidase Activity. Serum β-glucuronidase activity was increased significantly after the injection of endotoxin. The increase in the activity was inhibited by the pretreatment of 50 mg/kg of SOD, but could not be by catalase.

The Changes in Serum Cathepsin B Activity. Serum levels of cathepsin B was increased significantly after the injection of endotoxin. The increase in the level was inhibited by the administration of 50 mg/kg of SOD or 1.0 mg/kg of catalase.

The Changes in TBA Reactants. Serum TBA reactants were increased significantly 45 min after the injection of endotoxin. The increase was inhibited by the administration of SOD but not by catalase.

Discussion

The present study demonstrated the protective effects of SOD and catalase against the aggravation of DIC and shock in rats induced by endotoxin. These findings suggest that superoxide and hydrogen peroxide can affect the experimental DIC and shock states.

Hydrogen peroxide and superoxide radicals have been reported to cause aggregation of blood platelets [18, 19]. Platelet aggregation has been shown to be inhibited by a variety of antioxidants and by scavengers of hydroxyl radicals and singlet oxygen [20–23]. These findings suggest a role of reactive oxygen species in platelet activation.

Superoxide radicals are commonplace products of the biological reduction of oxygen [1, 2]. Fluxes of superoxide radicals have been shown to induce lipid peroxidation [24], damage membranes [25, 26], and kill cells [27]. Lipid peroxidation mediated by active oxygens is believed to be one of the important causes of cell membrane destruction and cell damage [28]. The injury to the endothelial cells, platelets or any kind of tissues may activate the coagulation system [29]. Our findings and those of other investigators suggest that superoxide radicals and hydrogen peroxide may play some role in the thrombus formation in the experimental DIC induced by endotoxin.

Novelli and De Gaudio [30] suggested that oxygen free radicals could be implicated in the pathogenesis of shock states. The radicals generation are increased in some critical states directly related to shock as tissue hypoxia [31], incomplete ischemia and tissue reperfusion [32], activation of arachidonic acid cascade [33] and complement-induced granulocyte aggregation [34].

SOD and catalase can prevent the aggravation of endotoxin shock. *Gray* [7] was reported that this enzyme SOD enhances survival in endotoxin-challenged mice. These observations support the hypothesis that oxygen radicals can be produced in endotoxin shock. Serum TBA reactants were increased in DIC and shock states. We reported that the levels of TBA reactant, important and damaging products of free radical lipid peroxidation in abdominal aortic wall and ileum mucosa were increased in rats infused with endotoxin [35]. In shock states, these peroxidations may occur in several organs. Since the effectiveness of SOD and catalase against DIC and shock, these enzymes may be useful for the treatment of these pathological conditions.

References

1   Fridovich, I.: The biology of oxygen radicals. Science *201:* 875–880 (1978).
2   Fridovich, I.: Superoxide dismutase. A. Rev. Biochem. *44:* 147–159 (1975).
3   Sies, H.: Biochemie des Peroxysoms in der Leberzelle. Angew. Chem. *86:* 789–801 (1974).
4   Bielsky, B. H. J.; Allen, A. O.: Mechanism of disproportionation of superoxide radicals. J. phy. Chem., Ithaca *81:* 1048–1052 (1977).
5   Ogawa, R.; Morita, T.; Kunimoto, F.; Fujita, T.: Changes in hepatic lipoperoxide concentration in endotoxemic rats. Circul. Shock *9:* 369–374 (1982).
6   Johnston, R. B., Jr.; Guthrie, L. A.; McPhail, L. C: Priming of neutrophils for en-

hanced oxidative metabolism by bacterial endotoxin; in Greenwald, Cohen, Oxyradicals and their scavenger system, vol II, pp 69–73 (Elsevier, New York 1983).

7   Gray, B.: Effects of superoxide dismutase of lipopolysaccharide-stressed mice and alteration of lung enzyme levels by endotoxin. Toxicol. appl. Pharmacol. *60:* 479–484 (1981).

8   Ferreira, H. C.; Murat, L. G.: Immunological method for demonstrating fibrin degradation products in serum and its use the diagnosis of fibrinolytic states. Br. J. Haematol. *9:* 299–310 (1963).

9   Ratnoff, O. D.; Menzie, C. A.: A new method for the determination of fibrinogen in small samples of plasma. J. Lab. clin.. Med. *37:* 316–320 (1951).

10 Quick, A. J.; Stanly, B. M.; Bancroft, F. W.: A study of the coagulation defect in gemophilic and in jaundice. Am. J. med. Sci. *190:* 501–511 (1935).

11 Langdell, R. D.; Wagner, R. H.; Brinkhous, K. M.: Effect of antihemophilic factor on one-stage clotting time. J. Lab. clin. Med. *47:* 637–647 (1953).

12 Brecker, G.; Cronkite, E. P.: Morphology and enumeration of human blood platelets. J. appl. Physiol. *3:* 365–377 (1950).

13 Weigert, C.: Über eine neue Methode zur Färbung von Fibrin und von Microorganismen. Fortschr. Med. *5:* 228–232 (1887).

14 Bunag, R.: Validation in awake rats of a tail-cuff method for measuring systolic pressure. J. appl. Physiol. *34:* 279–282 (1973).

15 Andersch, M. A.; Szczypinski, A. J.: Use of *p*-nitrophenylphosphate as the substrate in determination of serum acid phosphatase. Am. J. clin. Path. *17:* 571–574 (1947).

16 Fishman, W. H.; Kato, K.; Anstiss, C. L.; Green, S.: Human serum β-glucuronidase; its measurement and some of its properties. Clinica chim. Acta *15:* 435–447 (1967).

17 Otto, K.; Bhakdi, S.: Zur Kenntnis des Kathepsins B: Spezifität und Eigenschaften. Hoppe-Seyler's Z. physiol. Chem. *350:* 1577–1588 (1969).

18 Handin, R. I.; Carabin, R.; Boxer, G. J.: Enhancement of platelet function by superoxide anion. J. clin. Invest. *59:* 959–965 (1977).

19 Higashi, O.; Kikuchi, Y.: Effects of vitamin E on the aggregation and the lipid peroxidation of platelets exposed to hydrogen peroxide. Tohoku J. exp. Med. *112:* 271–278 (1974).

20 Steiner, M.; Anastasi, J.: Vitamin E. An inhibitor of the platelet release reaction. J. clin. Invest. *57:* 732–737 (1976).

21 Panganamala, R. V.; Miller, J. S.; Gwebu, E. T.; Sharma, H. M.; Cornwell, D. G.: Differential inhibitory effects of vitamin E and other antioxidants on prostaglandin synthetase, platelet aggregation and lipoxidase. Prostaglandins *14:* 261–271 (1977).

22 White, J. G.; Rao, G. H. R.; Gerrard, J. M.: Effects of nitroblue tetrazolium and vitamin E on platelet ultrastructure aggregation, and secretion. Am. J. Path. *88:* 387–402 (1977).

23 Greenberg, M.; Grady, R. W.; Peterson, C. M.: Inhibition of platelet function with 2,3-dihydroxybenzoic acid. Br. J. Haemat. *37:* 569–577 (1977).

24 Kellogg, E. W., III; Fridovich, I.: Superoxide, hydrogen peroxide, and singlet oxygen in lipid peroxidation by a xanthine oxidase system. J. biol. Chem. *250:* 8812–8817 (1975).

25 Goldberg, B.; Stern, A.: Superoxide anion as a mediator of drug-induced oxidative hemolysis. J. biol. Chem. *251:* 6468–6470 (1976).

26 Kellogg, E. W., III; Fridovich, I.: Liposome oxidation and erythrocyte lysis by enzymically generated superoxide and hydrogen peroxide. J. biol. Chem. *252:* 6721–6728 (1977).

27 Michelson, A. M.; Buckingham, M. E.: Effects of superoxide radicals on myoblast growth and differentiation. Biochem. biophys. Res. Commun. *58:* 1079–1086 (1974).

28 Tappel. A. L.: Lipid peroxidation damage to cell components. Fed. Proc. *32:* 1870–1874 (1973).

29 Minna, J. D.; Robboy, S.; Colman, R. W.: Disseminated intravascular coagulation in man, pp. 3–18 (Thomas, Springfield 1974).

30 Novelli, G. P.; De Gaudio, A. R.: Oxygen free radicals in shock states; in Lewis, Haglund, Shock research, pp. 31–42 (Elsevier, Amsterdam 1983).

31 Demopoulos, H. B.; Flamm, E. S.; Pietronigro, D. D.; Seligman, M. L.: Free radical pathology and the microcirculation in the major central nervous system disorders. Acta physiol. scand. suppl. 492, pp. 91–119 (1980).

32 Del Maestro, R. F.; Thaw, H. H.; Biörk, J.; Planker, M.; Arfors, K. E.: Free radicals as mediators of tissue injury. Acta physiol. scand. suppl. 492, pp. 43–57 (1980).

33 Lefer, A. M.; Araki, H.; Okamatsu, S.: Beneficial actions of a free radical scavenger in traumatic shock and myocardial ischemia. Circul. Shock. *8:* 273–282 (1981).

34 Goldstein, I. M.; Roos, D.; Kaplan, H. B.; Weissmann, G.: Complement and immunoglobulins stimulate superoxide production by human leukocytes independently of phagocytosis. J. clin. Invest. *56:* 1155–1163 (1975).

35 Yoshikawa, T.; Murakami, M.; Furukawa, Y.; Kato, H.; Takemura, S.; Kondo, M.: Lipid peroxidation and experimental disseminated intravascular coagulation in rats induced by endotoxin. Thromb. Haemostasis *49:* 214–216 (1983).

Toshikazu Yoshikawa, MD, PhD, First Department of Medicine,
Kyoto Prefectural University of Medicine, Kamiko-ku, Kyoto 602 (Japan)

Novelli, Ursini (eds.), Oxygen Free Radicals in Shock. Int. Workshop,
Florence 1985, pp. 137–148 (Karger, Basel 1986)

Role of Vitamin E on Endotoxin-Induced Disseminated Intravascular Coagulation

T. Yoshikawa, M. Murakami, T. Tanigawa, S. Sugino, M. Kondo

First Department of Medicine, Kyoto Prefectural University of Medicine,
Kamikyo-ku, Kyoto, Japan

Disseminated intravascular coagulation (DIC) is a pathological syndrome in which the formation of fibrin thrombi, the consumption of specific plasma proteins, the loss of platelets, and the activation of the fibrinolytic system suggest the presence of thrombin in the systemic circulation. Clinically these effects can become manifest as diffuse hemorrhage and, less frequently, as thrombosis [1].

There is a close relationship between vitamin E and the coagulation system. Inhibition of platelet aggregation by vitamin E has been observed by several investigators [2–5]. Prolongation of plasma clotting time has been reported with vitamin E therapy, together with a decrease in the expected increment in blood fibrinolytic activity which follows venous occlusion [6]. An antithrombin effect of vitamin E has been reported by *Kay* et al. [7] but not confirmed [8].

The pathological lesions observed in porcine spontaneous deaths associated with vitamin E and selenium deficiency appear to be a manifestation of DIC, being characterized by multiple microthrombi in small vessels [9]. The generalized Shwartzman reaction, an experimental model of DIC, can be induced in pregnant rats by a vitamin E-deficient diet [10], and vitamin E-deficient pigs possess an enhanced susceptibility to this reaction with myocardial vascular lesions similar to those observed with the dietetic microangiopathy syndrome [11]. These reports suggest a relationship between vitamin E deficiency and DIC.

We reported that experimental DIC could be induced by a 4-hour sustained infusion of endotoxin in rats [12]. The present work was undertaken to study the preventive effect of vitamin E against DIC and to investigate the influence of vitamin E deficiency in these states.

Methods

Animals. Female Wistar rats (Keari Co., Ltd., Osaka, Japan) at 4 weeks of age were housed in groups of 4–5 in plastic cages with wood-chip bedding. The animal housing room was maintained at a relatively constant temperature (21 ± 3 °C) and humidity ($50 \pm 20\%$) and had a 12-hour light-dark cycle (lights on 0800–2000 hours). The rats were maintained on a standard diet, or a vitamin E-deficient or a vitamin E-supplemented diet for 4 months and given water ad libitum.

Experimental DIC. Experimental DIC was induced by a sustained infusion of 13.3 or 100 mg/kg of endotoxin, diluted in 11.4 ml of saline, into the femoral vein for 4 h with a syringe. For i. v. infusions the femoral vein of the rats were cannulated under light Nembutal (5 mg/100 g, i. p.) anesthesia.

To examine the preventive effect of vitamin E against the experimental DIC, the rats were injected with *DL*-α-tocopheryl acetate (Eisai Co., Tokyo, Japan) at 0.01, 0.1, 1.0 or 10.0 mg/kg/day intraperitoneally for 4 successive days; 10 min after the final injection, they were infused continuously with 100 mg/kg of endotoxin for 4 h. *DL*-α-tocopheryl acetate was dissolved in HCO-60 (polyethylene 60-hydrogenated castor oil, Eisai Co., Tokyo, Japan) at a concentration of 50 mg/ml.

Chemicals. Endotoxin *(Escherichia coli* 055: B5 lipopolysaccharide B, Difco Lab., Detroit, Mich.) was dissolved in pyrogen-free physiological saline before every experiment. All other chemicals were of reagent grade.

Diet. Basal vitamin E-deficient diet (Oriental Yeast Co., Ltd., Tokyo, Japan) contains 38.0% corn starch, 10.0% α-malt starch, 5.0% granulated sugar, 25.0% vitamin-free casein, 6.0% purified lard, 2.0% vitamin E mixture, 6.0% mineral mixture, and 8.0% powdered filter paper.

The vitamin mixture in 100 g basal diet contains 1,000 IU vitamin A acetate, 200 IU vitamin D_3, 2.4 mg thiamine HCl, 8.0 mg riboflavin, 1.6 mg pyridoxine HCl, 1.0 µg cyaocobalamin, 60.0 mg *L*-ascorbic acid, 10.4 mg vitamin K_3, 0.04 mg *D*-biotin, 0.4 mg folic acid, 10.0 mg Ca panthothenate, 10.0 mg *p*-aminobenzoic acid, 12.0 mg niacin, 12.0 mg inositol, and 400 mg choline. The vitamin E content of the vitamin E-deficient diet is 0.16 mg/100 g diet, and this amount is about 1.6% of normal diet. The vitamin E-supplemented diet was prepared by adding 20.0 mg of *DL*-α-tocopheryl acetate to 100 g basal vitamin E-deficient diet. The salt mixture in 100 g basal diet contains 692 mg K, 411 mg Ca, 270 mg Na, 86 mg Mg, 41 mg Fe, 0.4 mg Zn, 1.3 mg Mn, 0.08 mg Cu, and 7.7 mg I.

Blood Collection. Blood samples were withdrawn from the abdominal aorta into plastic syringes. The samples used for blood counts, platelet counts, and hematocrit were kept from clotting with 0.01 M sodium ethylenediaminetetraacetic acid, and those for prothrombin time (PT), partial thromboplastin time (PTT), and fibrinogen determination were diluted (1:9 vol) with 0.1 M sodium citrate.

Determination. The serum α-tocopherol content was determined by the method of *Abe* et al. [13]. Erythrocyte hemolysis against dialuric acid was measured according to the method of *Friedman* et al. [14]. Fibrinogen and fibrin degradation products (FDP) were determined by the latex agglutination test (Teikoku Hormone Mfg. Co., Tokyo, Japan) and expressed in micrograms per milliliter equivalent to human fibrinogen base [15]. Fibrinogen, PT, and PTT were measured according to the methods of *Ratnoff and Menzie* [16], *Quick* et al. [17], and *Langdell* et al. [18], respectively. Platelets were counted by phase-contrast microscopy [19]. Red and white blood cells were counted in an automatic blood cell counter (S-plus-II, Coulter Electronics Inc., Hialeah, Fla.). Hematocrit was determined in duplicate in microhematocrit tubes (Propper Mfg. Co., Inc., Long Island, N.Y.) centrifuged at 10,000 g for 6 min. Kidney tissue specimens were examined histologically by hematoxylin-eosin stain; 100 glomeruli were counted, and those having fibrin thrombi were expressed in percentages (percentage glomerular fibrin deposits, %GFD).

Results

Effects of Vitamin E on Experimental DIC

It was shown that DIC could be induced by 4 h sustained infusion of endotoxin in a dose of 100 mg/kg. Then this experimental model was used for studies on the protective effect of vitamin E at various doses against DIC. Before the infusion of endotoxin, 0.01, 0.1, 1.0 or 10.0 mg/kg/day of α-tocopheryl acetate was injected intraperitoneally for 4 successive days, and the preventive effect against endotoxin-induced DIC was examined (table I).

FDP. The FDP level was 2.3 ± 0.5 µg/ml ($\bar{x}$ ± SD) in normal rats infused continuously with physiological saline alone. The level increased to 43.1 ± 5.4 µg/ml in animals pretreated with 0.2 ml/kg/day of HCO-60, a solvent of α-tocopheryl acetate, for 4 successive days intraperitoneally, and given 100 mg/kg/4 h of endotoxin. When 1.0 or 10.0 mg/kg/day of α-tocopheryl acetate was administered to rats for 4 successive days before the infusion of endotoxin, the FDP level was reduced significantly.

Fibrinogen. The level of fibrinogen was 1.21 ± 0.30 g/l ($\bar{x}$ ± SD) in normal rats administered saline alone. The level decreased to 0.20 g/l or less in rats pretreated with 0.2 ml/kg/day of HCO–60 for successive days and infused with 100 mg/kg/4 h of endotoxin. The decrease of fibrinogen level was prevented significantly in rats administered with 0,1, 1.0 or 10.0 mg/kg/day of α-tocopheryl acetate for 4 successive days.

PT. The effect of α-tocopheryl acetate was examined in the same manner on the prolongation of PT in the case of experimental DIC. PT was 11.9 ± 0.8 s ($\bar{x}$ ± SD) in the normal rats infused with physiological saline solution for 4 h. It was prolonged to 22.9 ± 3.8 in rats pretreated with 0.2 ml/kg/day of HCO-60 for 4 successive days and infused with 100 mg/kg/4 h of endotoxin. The prolongation of PT was prevented significantly in rats administered with 1.0 or 10.0 mg/kg/day of α-toco-

Table I. Effect of vitamin E (α-tocopheryl acetate) on experimental disseminated intravascular coagulation in rats infused with 100 mg/kg of endotoxin for 4 h

	FDP µg/ml	Fibrinogen g/l	PT s	PTT s	Platelet count $\times 10^9/l$	GFD[a] %	N[b]
HCO-60[c]; 0.2 ml/kg (control)	43.1±5.4	< 0.2	22.9±3.8	121.6±12.4	235±63	75.2±9.7	12
α-Tocopheryl acetate							
0.01 mg/kg	40.7±5.9	0.21±0.04	21.9±3.7	102.1±16.4[f]	238±66	61.1±12.5[f]	10
0.1	38.6±5.4	0.25±0.05[f]	21.3±3.4	92.3±20.4[g]	301±61[e]	30.6±8.1[g]	10
1.0	5.5±0.5[g]	0.32±0.06[g]	17.6±3.7[f]	67.8±12.0[g]	559±101[g]	4.9±1.4[g]	10
10.0	5.1±1.0[g]	0.37±0.10[g]	15.1±2.3[g]	66.6±12.4[g]	662±111[g]	4.2±1.5[g]	10
Normal rats[d]	2.3±0.5[g]	1.21±0.30[g]	11.9±0.8[g]	59.8±5.2[g]	1076±281[g]	0[g]	14

Results are expressed as $\bar{x} \pm SD$.
[a] Glomerular fibrin deposits.
[b] Number of subjects.
[c] HCO-60 was injected intraperitoneally for 4 successive days before the infusion of endotoxin.
[d] Normal rats group was infused 11.4 ml/4 h of saline and treated with neither α-tocopheryl acetate nor endotoxin.
[e] $p < 0.05$.
[f] $p < 0.01$.
[g] $p < 0.001$ for difference to the values of rats pretreated with HCO-60 as controls by Student's t-test.

pheryl acetate for 4 successive days, as compared with the control rats pretreated with HCO-60 and infused with endotoxin.

PTT. PTT was 59.8 ± 5.2 s ($\bar{x} \pm SD$) in the normal rats infused physiological saline for 4 h. It was prolonged to 121.6 ± 12.4 s in rats pretreated with 0.2 ml/kg/day of HCO-60 for 4 successive days and infused with 100 mg/kg/4 h of endotoxin. Pretreatment of α-tocopheryl acetate at 0.01, 0.1, 1.0, or 10.0 mg/kg/day for 4 successive days inhibited the prolongation significantly.

Platelet Count. The platelet count was 1,076 ± 281× $10^9/l$ ($\bar{x} \pm SD$) in the normal rats group infused saline for 4 h. It was reduced to 235 ± 63 × $10^9/l$ in rats pretreated with 0.2 ml/kg/day of HCO-60 for 4 successive days and infused with 100 mg/kg/4 h of endotoxin. When rats were administered with 0.1, 1.0 or 10.0 mg/kg/day of α-tocopheryl acetate for 4 successive days, the reduction of platelet count was prevented significantly in these animals compared with those in the control rats pretreated with HCO-60 and infused with endotoxin.

Formation of Fibrin Thrombi in the Renal Glomeruli. Histological examination was made on the kidney collected from rats infused with 100 mg/kg/4 h of endotoxin. In it, 100 glomeruli were counted to detect those containing fibrin thrombi. The results obtained was expressed as %GFD. In the normal rats group infused with physiological saline solution for 4 h, %GFD were 0%; that is no fibrin thrombi were formed at all in any glomerulus. When rats were pretreated with 0.2 ml/kg/day of HCO-60 for 4 successive days and infused with 100 mg/kg/4 h of endotoxin, %GFD were as high as 75.2 ± 9.7%, that is, the renal glomeruli contained fibrin thrombi formed were highly observed. However, when rats were pretreated with 0.01, 0.1, 1.0 or 10.0 mg/kg/day of α-tocopheryl acetate for 4 successive days and infused with 100 mg/kg/4 h of endotoxin, the formation of fibrin thrombi was prevented significantly.

Direct Effect of α-Tocopheryl Acetate or HCO-60 on Various Parameters. Investigation was made to determine whether α-tocopheryl acetate or HCO-60 exerted direct effects on the measured values of various parameters. Rats were pretreated by i.p. injection with 100.0, 10.0, 1.0, 0.1 or 0.01 mg/kg/day of α-tocopheryl acetate, or 0.2 ml/kg/day of HCO-60 for 4 successive days. At 1, 2, 3, and 4 h after the final injection of α-tocopheryl acetate or HCO-60, FDP, fibrinogen, PT, PTT, platelet count and the number of renal glomeruli with fibrin thrombi were compared with the values obtained from rats receiving physiological saline as a control. There was no noticeable change.

Vitamin E Deficiency

The rats were maintained on a vitamin E-deficient diet for 4 months. The serum level of vitamin E (α-tocopherol) decreased to 1.9 ± 0.5 µg/ml ($\bar{x} \pm$ SD, n = 8) at 4 months of feeding. When the rate of erythrocyte hemolysis was determined with dialuric acid, the rate rose to $93.3 \pm 2.5\%$ ($\bar{x} \pm$ SD, n = 8) at 4 months of feeding with the vitamin E-deficient diet. This value, when compared to that of $2.1 \pm 1.6\%$ ($\bar{x} \pm$ SD, n = 8) obtained in rats maintained on an ordinary diet, clearly indicates that complete vitamin E deficiency was induced in the rats by a 4-month feeding on the vitamin E-deficient diet.

Vitamin E Supplementation

The rats maintained on a vitamin E-supplemented diet showed serum α-tocopherol levels of 20.4 ± 2.1 µg/ml ($\bar{x} \pm$ SD, n = 8) at 4

Table II. Fibrinogen and fibrin degradation product (FDP) levels[a] in rats deficient in or supplemented with vitamin E after 4-hour sustained infusion of 13.3 mg/kg of endotoxin or 11.4 ml of saline

	Saline µg/ml	Endotoxin µg/ml	p-value[b]
Vitamin E-deficient group	4.1±0.6 (n = 9)	13.8±6.1 (n = 19)	< 0.001
Vitamin E-supplemented group	3.9±0.5 (n = 8)	10.1±5.2 (n = 18)	< 0.05
p-value[c]	n. s.[d]	n. s.	

[a] FDP levels are expressed as x̄±SD.
[b,c] Multiple comparisons by Scheffe type [b]between saline- and endotoxin-induced rats and [c]between vitamin E-deficient and supplemented groups.
[d] Not significant.

Table III. Fibrinogen levels[a] in rats deficient in or supplemented with vitamin E after 4-hour sustained infusion of 13.3 mg/kg of endotoxin or 11.4 ml of saline

	Saline, g/l	Endotoxin, g/l	p-value[b]
Vitamin E-deficient group	1.38±0.27 (n = 9)	0.37±0.16 (n = 19)	< 0.001
Vitamin E-supplemented group	1.40±0.30 (n = 8)	0.62±0.11 (n = 18)	< 0.001
p-value[c]	n. s.[d]	< 0.001	

[a] Fibrinogen levels are expressed as x̄±SD.
[b,c] Multiple comparisons by Scheffe type [b]between saline- and endotoxin-induced rats and [c]between vitamin E-deficient and supplemented groups.
[d] Not significant.

months of feeding. The rate of erythrocyte hemolysis due to dialuric acid was $0.9 \pm 0.8\%$ ($\bar{x} \pm$ SD, n = 8).

Endotoxin-Induced DIC in Vitamin E-Deficient or Supplemented Rats
Hematocrit and Blood Cells. Hematocrit levels of rats deficient in or supplemented with vitamin E were $50.6 \pm 1.7\%$ ($\bar{x} \pm$ SD, n = 18) or $51.2 \pm 2.2\%$ (n = 16), respectively. The levels decreased to $43.8 \pm 3.2\%$ (n = 9) or $43.1 \pm 2.5\%$ (n = 8) after the infusion of 11.4 ml/rat/4 h of sa-

Table IV. Prothrombin time[a] in rats deficient in or supplemented with vitamin E after 4-hour sustained infusion of 13.3 mg/kg of endotoxin or 11.4 ml of saline

	Saline, s	Endotoxin, s	p-value[b]
Vitamin E-deficient group	11.7±0.8	15.7±2.3	< 0.001
	(n = 9)	(n = 19)	
Vitamin E-supplemented group	11.4±0.6	13.6±1.2	< 0.01
	(n = 8)	(n = 18)	
p-value[c]	n.s.[d]	< 0.001	

[a] Prothrombin time is expressed as $\bar{x}\pm SD$.
[b,c] Multiple comparisons by Scheffe type [b] between saline- and endotoxin-induced rats and [c] between vitamin E-deficient and supplemented groups.
[d] Not significant.

Table V. Partial thromboplastin time[a] in rats deficient in or supplemented with vitamin E after 4-hour sustained infusion of 13.3 mg/kg of endotoxin or 11.4 ml of saline

	Saline, s	Endotoxin, s	p-value[b]
Vitamin E-deficient group	38.9±5.1	75.0±20.3	< 0.001
	(n = 9)	(n = 19)	
Vitamin E-supplemented group	38.7±6.7	51.7±10.3	n.s.
	(n = 8)	(n = 18)	
p-value[c]	n.s.[d]	< 0.001	

[a] Partial thromboplastin time is expressed as $\bar{x}\pm SD$.
[b,c] Multiple comparisons by Scheffe type [b] between saline- and endotoxin-induced rats and [c] between vitamin E-deficient and supplemented groups.
[d] Not significant.

line, and 42.9 ± 3.3% (n = 8) or 43.3 ± 3.6% (n = 8) after the infusion of 13.3 mg/kg/4 h of endotoxin, respectively. Red and white blood cell counts decreased only slightly or not at all after the treatment of saline or endotoxin.

FDP. When rats were infused with 13.3 mg/kg/4 h of endotoxin, diluted in 11.4 ml of physiological saline, the FDP level increased significantly in both vitamin E-deficient and supplemented rats. There was no significant difference in the increase between the two groups (table II).

Table VI. Platelet counts[a] in rats deficient in or supplemented with vitamin E after 4-hour sustained infusion of 13.3 mg/kg of endotoxin or 11.4 ml of saline

	Saline, $\times 10^9/l$	Endotoxin, $\times 10^9/l$	p-value[b]
Vitamin E-deficient group	605 ± 78 (n = 9)	256 ± 94 (n = 19)	< 0.001
Vitamin E-supplemented group	611 ± 88 (n = 8)	348 ± 102 (n = 18)	< 0.001
p-value[c]	n.s.[d]	< 0.01	

[a] Platelet counts are expressed as $\bar{x}\pm SD$.
[b,c] Multiple comparisons by Scheffe type [b]between saline- and endotoxin-induced rats and [c]between vitamin E-deficient and supplemented groups.
[d] Not significant.

Table VII. Percentage glomerular fibrin deposits[a] in rats deficient in or supplemented with vitamin E after 4-hour sustained infusion of 13.3 mg/kg of endotoxin or 11.4 ml of saline

	Saline, %	Endotoxin, %
Vitamin E-deficient group	0 (n = 9)	57.3 ± 9.3 (n = 19)
Vitamin E-supplemented group	0 (n = 8)	40.2 ± 9.1 (n = 18)
p-value[b]		< 0.001

[a] Percentage glomerular fibrin deposits are expressed as $\bar{x}\pm SD$.
[b] Multiple comparisons by Scheffe type between vitamin E-deficient and supplemented groups.

Fibrinogen. The fibrinogen level decreased in vitamin E-deficient rats infused with endotoxin. The decrease in fibrinogen level of vitamin E-supplemented rats was less than that of the vitamin E-deficient group (p < 0.001) (table III).

PT. PT was prolonged in vitamin E-deficient rats infused with endotoxin. The prolongation in vitamin E-supplemented rats was less than that in vitamin E-deficient rats (table IV).

PTT. Although PTT was prolonged significantly in vitamin E-deficient rats after the infusion of endotoxin, the prolongation was slight in vitamin E-supplemented rats (table V).

Platelet Counts. The platelet counts were decreased in vitamin E-deficient rats infused with endotoxin. The reduction in platelet count was less in vitamin E-supplemented rats than in vitamin E-deficient (table VI).

Formation of Fibrin Thrombi in the Renal Glomeruli. In the histological examination of kidney specimens collected from rats infused with endotoxin, glomeruli were counted to detect those containing fibrin thrombi. In both groups of rats infused with physiological saline solution for 4 h the %GFD was 0%; that is, no fibrin thrombi were formed in any of the glomeruli. When the rats were infused with endotoxin %GFD was increased. However, the formation of fibrin thrombi was less in vitamin E-supplemented rats than in vitamin E-deficient rats (table VII).

Discussion

Vitamin E, first identified by *Evans and Bishop* [20] in 1922, is a required nutrient which has been the focus of new interest since clinical trials suggested it might be effective in reducing the incidence of certain thrombotic disorders, including intermittent claudication [21], cerebral arteriosclerosis [22], and coronary artery disease [23].

We had already reported that experimental model of DIC could be induced by 4 h sustained infusion of endotoxin in a dose of 100 mg/kg in rats. Using this experimental model of DIC, the investigation was made on the effect of vitamin E against DIC.

When compared with rats given HCO-60, a solvent of α-tocopheryl acetate, the significant prevention against DIC was noted in all the parameters in rats treated with 1.0 or 10.0 mg/kg/day of α-tocopheryl acetate for 4 successive days.

Clinically, these doses of α-tocopheryl acetate can be administered to man and it can be suggested that this agent may be useful for the prevention of DIC. However, there is no evidence that this agent is effective after the onset of DIC.

On the other hand, the changes in several coagulatory parameters of rats deficient in vitamin E were remarkable and only slight in those supplemented with vitamin E when rats were infused 13.3 mg/kg of endotoxin for 4 h.

It was indicated that a thrombotic state is induced by intake of a diet rich in polyunsaturated fats and deficient in vitamin E [24, 25].

Inhibition of platelet aggregation by vitamin E has been reported [2–5]. In addition, vitamin E was of preventive value in thromboembolism because of its antithrombin properties [26]. These findings concerning vitamin E might shed light on the possible mechanism of this agent in inhibiting the aggravation of endotoxin-induced experimental DIC.

Another possible mechanism for vitamin E to prevent the aggravation of DIC is through its lipid antioxidant activity, namely, inhibiting the peroxidation of polyunsaturated lipids [27]. Platelets may be induced to aggregate with hydrogen peroxide, and this aggregation can be prevented with tocopherols [2]. It appears that vitamin E is necessary to inhibit the peroxidation of unsaturated fatty acids which form an integral part of membrane structures [28]. We have reported that the levels of thiobarbituric acid reactive substances, important and damaging products of lipid peroxidation, were significantly increased in rats infused with endotoxin [29].

We recently reported that superoxide dismutase and catalase could prevent the aggregation of endotoxin-induced DIC in rats [30]. These findings suggest a role of reactive oxygen species in platelet activation and in the thrombus formation. Vitamin E can quench free radicals and active oxygens and reduce lipid peroxidation [31]. These effects of vitamin E may be beneficial for the prevention of DIC. However, other possible mechanisms of vitamin E in inhibiting DIC must be considered, because the agent has many biological effects.

References

1   Minna, J. D.; Robboy, S. I.; Colman, R. W.: Disseminated intravascular coagulation in man, pp. 3–18 (Thomas, Springfield 1974).

2   Higashi, O.; Kikuchi, Y.: Effects of vitamin E on the aggregation and the lipid peroxidation of platelets exposed to hydrogen peroxide. Tohoku J. exp. Med. *112:* 271–278 (1974).

3   Fong, J. S.: Alpha-tocopherol: its inhibition on human platelet aggregation. Experientia *32:* 639–641 (1976).

4   Machlin, L. J.; Filipski, R.; Willis, A. L.; Kuhn, D. C.: Influence of vitamin E on platelet aggregation and thrombocythemia in the rats. Proc. Soc. exp. Biol. Med. *149:* 275–277 (1975).

5   Steiner, M.: Inhibition of platelet aggregation by alpha-tocopherol; in de Duve, Hayaishi, Tocopherol, oxygen and biomembranes, pp. 143–163 (Elsevier/North-Holland, Amsterdam 1978).

6   Korsan-Bengsten, K.; Elmfeldt, D.; Holm, T.: Prolonged plasma clotting time and decreased fibrinolysis after long-term treatment with alphatocopherol. Thromb. Diath. haemorrh. *31:* 505–512 (1974).

7   Kay, J. H.; Hutton, S. B., Jr.; Weiss, G. N.; Ochsner, A.: Studies on an antithrombin; plasma antithrombin test for prediction of intravascular clotting. Surgery, St Louis *28:* 24–28 (1950).

8   Olson, R. E.: Vitamin E and its relation to heart disease. Circulation *48:* 179–184 (1973).

9   Fontaine, M.; Valli, V. E. O.; Young, L. G.: Studies on vitamin E and selenium deficiency in young pigs. IV. Effect on coagulation system. Can. J. comp. Med. *41:* 64–76 (1977).

10   Stamler, F. W.: Fatal eclamtic disease of pregnant rats fed anti-vitamin E stress diet. Am. J. Path. *35:* 1207–1231 (1959).

11   Teije, G.; Nordstoga, K.; Fjolstad, M.; Nafstadt, I.: The generalized Shwartzman reaction in pigs induced by diet and single injectin of disintegrated cells or partially purified endotoxin from *Escherichia coli.* Acta vet. scand. *14:* 92–106 (1973).

12   Yoshikawa, T.; Furukawa, Y.; Murakami, M.; Takemura, S.; Kondo, M.: Experimental model of disseminated intravascular coagulation induced by sustained infusion of endotoxin. Res. exp. Med. *179:* 223–228 (1981).

13   Abe, K.; Yuguchi, Y.; Katsui, G.: Quantitative determination of tocopherols by high-speed liquid chromatography. J. Nutr. Sci. Vitaminol. *21:* 183–188 (1975).

14   Friedman, L.; Weiss, W.; Wherry, F.; Line, O. L.: Bioassay of vitamin E by dialuric acid hemolysis method. J. Nutr. *65:* 143–160 (1958).

15   Ferreira, H. C.; Murat, L. G.: Immunological method for demonstrating fibrin degradation products in serum and its use the diagnosis of fibrinolytic states. Br. J. Haemat. *9:* 299–310 (1963).

16   Ratnoff, O. D.; Menzie, C.: A new method for the determination of fibrinogen in small samples of plasma. J. Lab. clin. Med. *37:* 316–320 (1951).

17   Quick, A. J.; Stanly, B. M.; Bancroft, F. W.: A study of the coagulation defect in hemophilic and in jaundice. Am. J. med. Sci. *190:* 501–511 (1935).

18   Langdell, R. D.; Wanger, R. H.; Brinkhous, K. M.: Effect of antihemophilic factor on one-stage clotting time. J. Lab. clin. Med. *47:* 637–647 (1953).

19   Brecker, G.; Cronkite, E. P.: Morphology and enumeration of human blood platelets. J. appl. Physiol. *3:* 365–377 (1950).

20   Evans, H. M.; Bishop, K. S.: On the existence of a hitherto unrecognized dietary factor essential for reproduction. Science *56:* 650–651 (1922).

21   Williams, H. T.; Fenna, D.; Macbeth, R. A.: Alpha-tocopherol in the treatment of intermittent claudication. Surgery Gynec. Obstet. *132:* 662–666 (1971).

22   Eisen, M. E.; Gross, H.: Vitamin E in arteriosclerotic and peripheral vascular disease. N. Y. med. J. *2422:* 49–53 (1949).

23   Toone, W. M.: Effects of vitamin E, good or bad. New Engl. J. Med. *289:* 979–980 (1973).

24   Nafstad, I.; Tollersrud, S.: The vitamin E-deficiency syndrome in pigs. I. Pathological changes. Acta vet. scand. *11:* 452–480 (1970).

25   Nafstad, I.: Endothelial changes and platelet thrombosis associated with PUFA-rich, vitamin E-deficient diet fed to pig. Thromb. Res. *5:* 251–258 (1974).

26   Ochsner, A.: Thromboembolism. New Engl. J. Med. *271:* 211 (1964).

27 Tappel, A. L.: Vitamin E and free radical peroxidation of lipids. Ann. N. Y. Acad. Sci. *203:* 12–28 (1972).
28 Strøm, E.; Nordøy, A.: Alpha-tocopherol (vitamin E) in human platelets. Thromb. Res. *4:* suppl., pp. 73–74 (1974).
29 Yoshikawa, T.; Murakami, M.; Furukawa, Y.; Kato, H.; Takemura, S.; Kondo, M.: Lipid peroxidation and experimental disseminated intravascular coagulation in rats induced by endotoxin. Thromb. Haemostasis *49:* 214–216 (1983).
30 Yoshikawa, T.; Murakami, M.; Yoshida, N.; Seto, O.; Kondo, M.: Effects of superoxide dismutase and catalase on disseminated intravascular coagulation in rats. Thromb. Haemostasis *50:* 869–872 (1983).
31 McCay, P. B.; King, M. M.: Vitamin E: its role as a biologic free radical scavenger and its relationship to the microsomal mixed-function oxidase system; in Machlin, Vitamin E: a comprehensive treatise, pp. 289–317 (Dekker, New York 1978).

T. Yoshikawa, MD, PhD , First Department of Medicine, Kyoto Prefectural University of Medicine, Kamikyo-ku, Kyoto 602 (Japan)

Novelli, Ursini (eds.), Oxygen Free Radicals in Shock. Int. Workshop,
Florence 1985, pp. 149–164 (Karger, Basel 1986)

Oxygen-Derived Free Radicals and Myocardial Injury: A Critical Role for Xanthine Oxidase?[1]

*David J. Hearse[a], Allan S. Manning[a], James M. Downey[b],
Derek M. Yellon[a]*

[a]The Heart Research Unit, The Rayne Institute, St. Thomas' Hospital, London,
England; [b]The Department of Physiology, University of Alabama, Mobile, Ala.,
USA

Introduction

Myocardial ischaemia initiates a complex sequence of progressively more severe cellular reactions which, if not checked, may lead to cell death and tissue necrosis. In the early stages of myocardial ischaemia cellular injury is reversible, such that if adequate coronary flow is restored any metabolic imbalance is rapidly normalized and the tissue will return to normal contractile function. As the duration or severity of ischaemia increases and the damage becomes more severe reperfusion may not result in an immediate return of function and recovery may be delayed for some time. For example, after 15 min of ischaemia adenine nucleotide content [1] and/or contractile activity [2] may be depressed for almost a week after the onset of reperfusion. However, as long as cellular injury is in the reversible phase a full recovery of function will ultimately occur. Eventually however ischaemic injury will become so severe that irreversible damage will occur and from this point on the tissue is destined to cell death and necrosis and reperfusion has no beneficial effect. The exact mechanism for the critical transition from reversible to irreversible injury remains unknown but a number of factors have been suggested as important determinants of cell death or survival; these include myocardial

[1] This work was supported in part by grants from the British Heart Foundation, The Medical Research Council of Great Britain and St Thomas' Hospital Research Endowments Fund. The assistance of Mrs *C. Erlebach* is gratefully acknowledged.

high energy phosphate depletion, intracellular calcium overload, loss of ionic homeostasis, labilization of lysosomes and phospholipase-mediated membrane injury. Recently, oxygen-derived free radicals such as superoxide (O_2^-) and the hydroxyl radical (OH^-), have been suggested [3–5] as important mediators of ischaemic injury.

Reperfusion of ischaemic tissue, either by surgical intervention or the natural growth of collaterals, is a necessary step if one wishes to halt the advance of ischaemic injury and promote tissue recovery. However, despite its obvious benefits reperfusion is not without hazard [6]. In moderately ischaemic tissue in which injury is in its reversible phase reperfusion can trigger potentially lethal arrhythmias such as ventricular fibrillation [7] and, in severely injured tissue, reperfusion can alter and accelerate the processes of cell death [8] and may lead to haemorrhage and other forms of severe tissue injury. As with ischaemia-induced injury, controversy exists over the mechanisms underlying various unfavourable reperfusion-induced changes. Calcium overload and cell swelling have received considerable attention and more recently free radicals have been proposed [7, 9, 10] as important contributors.

In discussing therapeutic interventions for the prevention of injury during ischaemia and reperfusion, two main objectives can be defined: firstly, to minimize the number of cells which become irreversibly injured and die (i.e. in the case of regional ischaemia reducing the size of an evolving infarct) and secondly, to prevent the occurrence of arrhythmias which may impair contractile performance or more importantly may be lethal. During the last decade there have been many studies of myocardial protection in which a wide variety of drugs and biochemical interventions have been claimed [11–14] to reduce infarct size and/or combat arrhythmias. To this list we can now add 'anti-free radical' agents such as allopurinol (which inhibits xanthine oxidase and hence limits free radical production during hypoxanthine degradation) and the enzyme superoxide dismutase (which along with a number of organic agents, such as glutathione, mannitol and ascorbate, can scavenge free radicals once they are formed). In 1971, *De Wall* et al. [15] reported that allopurinol treatment reduced arrhythmias in dog hearts with regional ischaemia. The anti-arrhythmic effects of allopurinol were originally [15] thought to reflect purine conservation in these hearts, but we would now suggest that allopurinol may act by suppressing superoxide production via xanthine oxidase. We, and others [9, 10], have now also shown that allopurinol and a variety of scavengers can effectively combat reperfusion-induced arrhythmias. In addition to the promising anti-arrhythmic properties of anti-free radical interventions, their potential clinical importance has recently been reinforced by the oberservation that, in a number of experimental models, infarct size can be considerably reduced by interventions, such as allopurinol [16], superoxide dismutase and catalase [17], and a variety of agents and procedures which limit leucocyte infiltration (an additional potential source or free radicals) [18] into ischaemic tissue.

The objective of the present article is to review some recent studies of the anti-infarct and anti-arrhythmic properties of anti-free radical interventions.

Free Radicals and Infarct Size

Since there is a direct relationship between mortality and the size of an evolving myocardial infarct considerable efforts have been made to develop interventions which prevent cell death and so reduce the size of an evolving infarct. In this section we will present a new closed chest dog model of coronary embolization and we will show how we have used this model (with its capacity for risk zone analysis) to assess whether the inhibition of xanthine oxidase with allopurinol is able to result in a sustained reduction of infarct size.

Dog Preparation for the Evaluation of Anti-Infarct Agents

In all studies adult greyhounds weighing between 19 and 32 kg were used, the dogs were anaesthetized with sodium pentabarbitone (30 mg/kg i.v.) and each received penicillin (1×10^6 U i.m.) to prevent postoperative infection. All experiments were carried out under sterile conditions.

The animals were intubated and a catheter was placed in the jugular vein for drug administration. The left carotid artery was exposed and a rigid cannula [19] with a pressure transducer attached to a side arm was inserted. The cannula was advanced into coronary ostium, successful insertion being confirmed by the pressure at the tip of the cannula decreasing from an aortic value to a distal coronary value. Coronary occlusion was then accomplished by administering a plastic bead (2.0 mm diameter) via the cannula. The cannula was then withdrawn from the coronary ostium and acvanced into the left ventricle. Approximately 2×10^6 radioactive microspheres (15 μm diameter), labelled with ^{141}Ce were then injected into the left ventricle. The cannula was then removed, the carotid artery cutdown was repaired and the dogs were returned to the recovery room. In this study all dogs received intravenous heparin (7,000 U every 8 h) to prevent clot formation proximal to the embolus.

Drug Administration. To evaluate the ability of allopurinol to salvage the myocardium at risk the dogs were divided into two groups. Firstly, there was an allopurinol

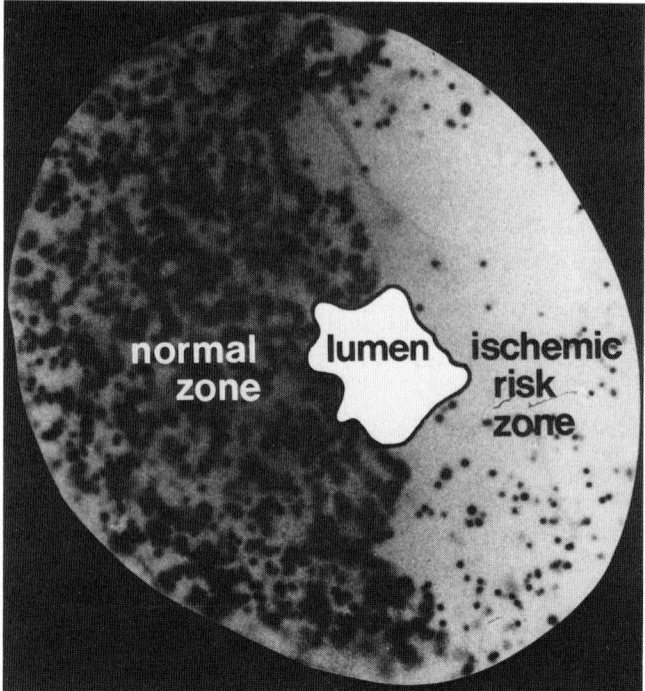

Fig. 1. Autoradiographic delineation of risk zone. An autoradiogram obtained from a slice of dog heart which had been subjected to coronary embolization and 24 h of ischaemia. Radioactive microspheres (^{141}Ce) administered at the onset of ischaemia are delivered to, and retained by, normally perfused tissue. The ischaemic zone (region at risk of infarction) is clearly defined by the lack of microspheres.

group (n = 10) where the dogs received 400 mg of allopurinol in their food 24 h before surgery and following anaesthesia a bolus (25 mg/kg) of allopurinol was administered intravenously over 10 min (this was repeated every 8 h). The second group of dogs (n = 10) received a 20-ml bolus of saline every 8 h.

Measurement of the Region at Risk and Infarct Size. After 24 h of coronary embolization the animals were re-anaesthetized and the hearts removed. The hearts were sliced into 5 mm thick sections from the apex to the base and were incubated in triphenyltetrazolium (TPT) for 20 min. At the end of the incubation period the slices were dipped in 10% formalin to fix the tissue and enhance the colour contrast. The tissue slices were then placed on an acrylic sheet, basal surface up, and covered with a layer of plastic film; this was followed by a sheet of X-ray film. Finally a second acrylic sheet was placed on top of the film and the two acrylic sheets were clamped tightly together. The assembly was put in a light-proof box at –20 °C for an appropriate exposure period (24–72 h depending upon the specific activity of the microspheres). The

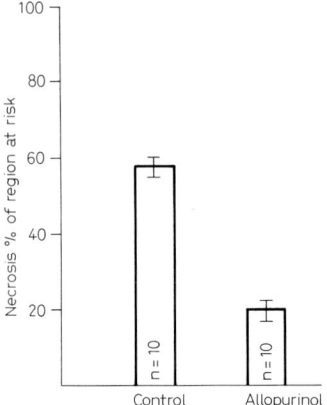

Fig. 2. Allopurinol and infarct size. Effect of allopurinol on the extent of tissue necrosis in the dog heart with 24 h of coronary artery embolization. Infarct size is expressed as a percent of the region at risk. Bars represent the SEM.

film was then removed and developed and figure 1 shows a typical autoradiogram. Each [141]Ce-labelled microsphere produces a sharp spot on the film. The autoradiogram thus reveals the area of normal blood flow around the field of the embolized artery and defines the area of underperfusion, i. e. the area at risk of infarction.

The areas of the basal surface of the infarct and the risk zone were measured by computerized planimetry and the volume of the infarct and the risk zone were calculated by multiplying the area by the thickness (5 mm) of each slice. The values of the sequential slices were then summed to give the total volume for each heart. The percentage of the region at risk which had infarcted could then be determined.

Allopurinol and Infarct Size

The results of this study are shown in figure 2. Infarct size in the drug-treated group was approximately 40% smaller than in the control group. The amount of the risk zone that infarcted in the drug group was $20.0 \pm 2.3\%$ whereas for the control group the value was $58.0 \pm 2.7\%$.

Possible Mechanisms for Allopurinol Protection. Ischaemia results in the net degradation of high energy phosphates, a process which terminates in the xanthine oxidase-mediated conversion of hypoxanthine to uric acid. Conversion to uric acid results in the irreversible loss of purine base from the nucleotide pool. It has been postulated that protection of the nucleotide pool could explain the beneficial effect of allopurinol during periods of oxygen depletion [15] and recent reports do indicate a beneficial effect when the purine base is replenished exogenously [20, 21]. For our preparation there are however two arguments against the nucleotide pool theory. Firstly, the unphosphorylated purines (adenosine, hypoxanthine and xanthine) are all lipid-soluble and can be washed out of the tissue. Since collateral flow in the ischaemic tissue in our preparation was in the order of 0.1–0.2 ml/min/g of tissue, this would result in a half-life for the unphosphorylated purines of less than 7 min. A blockade of xanthine oxidase alone would not be expected to keep the purines from being lost from the ischaemic zone. Secondly, there is little evidence that the loss of unphosphorylated purines represents a lethal event during ischaemia. In the absence of any other proposed mechanisms we would therefore suggest that the ability of allopurinol to limit infarct size in both the permanently ischaemic and the reperfused myocardium is most likely to be due directly to its ability to inhibit xanthine oxidase and thereby prevent the production of superoxide.

Applicability to Man?

Whether allopurinol treatment in the coronary patient will result in a comparable salvage of myocardium depends on several key factors. Firstly: do human hearts contain xanthine oxidase? There are few published studies, but at least one report [22] claims that the enzyme does exist. However, it is unknown what the kinetics are for conversion of the dehydrogenase to the oxidase form of the enzyme in the human heart.

A second key factor is how much, if any, of the region at risk in the human heart receives collateral blood flow in the range to qualify for tissue salvage? Our current findings would suggest that in the dog heart the threshold for natural necrosis is at a residual flow level of approximately 30% of normal or less (a value which may be somewhat lower in the presence of anti-infarct agents which slow cell death or inhibit unfavourable ischaemia-induced cellular changes). Does the coronary patient have a significant quantity of tissue in such a collateral flow category to justify the procedure? It is well-known that humans have fewer collateral vessels than do their canine counterparts [23]. On the basis of this one might be inclined to say no; however, collateral vessels are known to proliferate with the onset of coronary artery narrowing, and if the occlusion is gradual (as it usually is in the coronary patient) then collateralization can be quite extensive by the time that the disease becomes symptomatic. Recent studies [24] of autopsy material have suggested that a significant proportion of the risk zone in patients can undergo natural salvage, indi-

cating that not all of the tissue in the human risk zone is severely ischaemic and that a substantial quantity of myocardium in the average patient
may be receiving sufficient flow to make drug-mediated salvage a feasible proposition.

The final factor which requires consideration is the question of how
should a drug like allopurinol be administered? Allopurinol is very specific in its action, it is well-tolerated and it has few side-effects. Peak
blood levels are reached within 2 h of oral administration. Our current
health care system does not normally allow us to identify many coronary
patients earlier than 3 h postocclusion and our animal laboratory experi-
ence would indicate that much beyond this time relatively little tissue will
remain amenable to drug-mediated salvage. Postocclusion administration
of allopurinol, or any other drug, is therefore unlikely to have a major
impact upon infarct size in many patients. However, because of the nature of allopurinol, its low toxicity and its low cost, the drug could con-
ceivably be administered prophylactically to patients at high risk of infarction. Furthermore, in view of the large number of patients already
taking allopurinol for other indications, it may already be exerting an effect upon coronary heart disease statistics. Clearly, considerably more
work is required before the full potential of this drug, or any other anti-
free radical agent, as an anti-infarct agent can be assessed.

Free Radicals and Arrhythmias

As discussed at the beginning of this article, the two main consequences of ischaemia and reperfusion are cellular injury leading to cell
death and disorders of cardiac rhythm. In the latter instance arrhythmias
may lead to cell death in what is otherwise potentially viable tissue. The
mechanisms underlying the genesis and control of ischaemia- and reperfusion-induced arrhythmias have been reviewed extensively elsewhere [7,
25, 26]. Similarly detailed information is available to describe the various
types of arrhythmias which may arise (fibrillation, tachycardia, ectopics,
etc.), the different times at which they arise (early and late arrhythmias),
and the differences in origin and pharmacological control between reper-
fusion- and ischaemia-induced arrhythmias [7, 26]. Consideration of the
above-cited literature would indicate that a number of factors might be
responsible for the genesis of various arrhythmias; these include stimulation of α- and/or β-adrenergic receptors [27, 28]; elevation of intracellu-

lar adenosine $3':5'$-cyclic phosphate (cAMP) content [29]; the formation and release of lysophosphatides [30]; disturbances of ionic homeostasis (particularly transsarcolemmal shifts of potassium and calcium) [31, 32]; and the metabolism of free fatty acids [33, 34]. More recently it has been suggested that free radical production may play an important role [9, 10] and that inhibition of free radicals, particularly those derived via the xanthine oxidase pathway, might pave the way to a new and important therapeutic principle. In this section we will describe studies to assess the ability of allopurinol to influence the severity of ischaemia- and reperfusion-induced arrhythmias in the anaesthetized rat.

Rat Preparation for the Study of Ischaemia- and Reperfusion-Induced Arrhythmias

Surgical Procedure. Rats were anaesthetized with pentobarbitone, and were ventilated with room air. The electrocardiogram was recorded via standard limb leads. The chest was then opened via a left thoracotomy and, using a modification of the Selye technique [35], the heart was gently exteriorized by pressure on the abdomen. A ligature was then rapidly placed around the left anterior descending coronary artery, close to its origin, and the heart was replaced in the thoracic cavity with the ligature ends exteriorized.

For the study of arrhythmias induced by ischaemia, both ends of the ligature were passed through a small plastic tube which could then be pressed on the surface of the heart directly above the left anterior descending coronary artery. The resulting arterial occlusion could then be maintained for the required time by clamping the plastic tube and ligature. For the study of reperfusion-induced arrhythmias, the left anterior descending coronary artery was occluded in the same manner (for 5 min), reperfusion could be achieved by removing the clamp and tube.

Validation of Coronary Occlusion and Reperfusion. To ensure that occlusion and reperfusion had been adequately accomplished, Evans Blue dye was administered intravenously. If administered after a period of continuous coronary artery occlusion, the generation of an area of regional ischaemia could be confirmed by the presence of a non-stained area of myocardium. In the reperfusion experiments, to ensure that reperfusion had occurred, the dye was routinely administered at the end of the reperfusion period. A uniformly stained heart was taken as evidence that reperfusion had occurred. Further evidence of successful occlusion and reperfusion was always available from the electrocardiographic recording.

Evaluation of Rhythm Disturbances. High-speed electrocardiograms were analysed for the number of premature ventricular complexes (PVC), the incidence and duration of ventricular fibrillation (VF) and the incidence and duration of ventricular tachycardia (VT).

Allopurinol and Ischaemia- and Reperfusion-Induced Arrhythmias

Rats were treated with allopurinol (20 mg/kg body wt) orally 24 h prior to study. A second similar dose was given, this time intravenously, after the animal had been

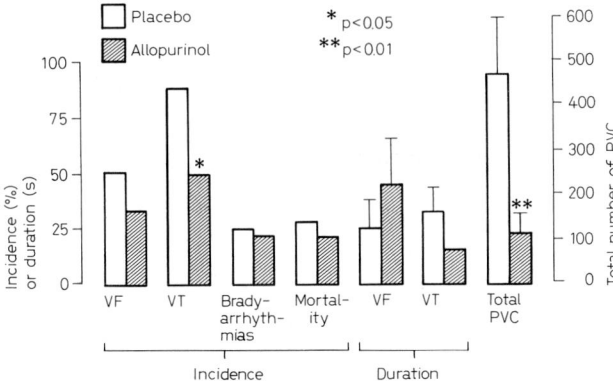

Fig. 3. Effect of allupurinol on ischaemia-induced arrhythmias in the in vivo anaesthetized rat. Rats were subjected to 30 min of regional ischaemia during which time myocardial rhythm was continuously monitored. VF = Ventricular fibrillation; VT = ventricular tachycardia; PVC = premature ventricular complexes. Values of duration of VF and VT are shown as the mean ± SEM; n = 18 animals in each group.

anaesthetized, 15 min prior to coronary occlusion. The control rats received equivalent amounts of saline, adjusted to the same pH as the suspension of allopurinol administered.

Allopurinol and Ischaemia-Induced Arrhythmias. In placebo-treated control animals (n = 18) severe ventricular arrhythmias began after 4–5 min of coronary artery occlusion; these peaked after 10 min and then normally subsided by 18–20 min. The mean time to onset of ventricular fibrillation was 303±101 s and the mean time to onset of VT was 344±32 s. As shown in figure 3, 50% of animals exhibited ventricular fibrillation and almost 90% exhibited ventricular tachycardia.

Allopurinol treatment had no significant effect on the incidence of ventricular fibrillation, the duration of ventricular fibrillation, the duration of ventricular tachycardia, and mortality. However, the incidence of ventricular tachycardia was reduced from 89 to 50% (p < 0.05) and the total number of premature ventricular complexes (in those animals surviving the 30 min period of coronary artery occlusion) was reduced from 471±128 to 116±46 (p < 0.02).

Allopurinol and Reperfusion-Induced Arrhythmias. We and others have shown that, in a number of species, vulnerability to reperfusion-induced arrhythmias is critically dependent upon the duration of the pre-

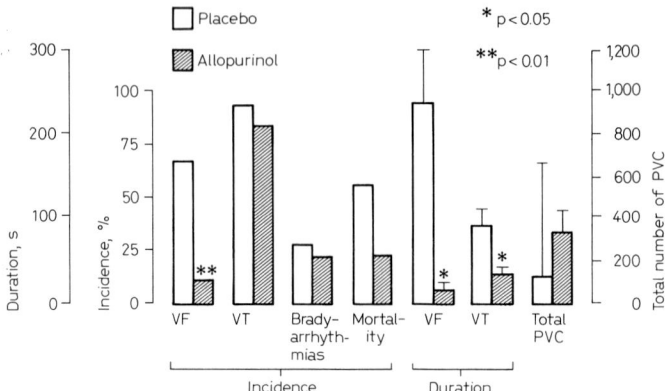

Fig. 4. Effect of allopurinol on reperfusion-induced arrhythmias in the in vivo anaesthetized rat. Rats were subjected to 5 min of regional ischaemia followed by reperfusion during which time cardiac rhythm was monitored. Abbrevations are as in figure 3. Values of duration of VF and VT are shown as the mean ± SEM; n = 18 animals in each group.

ceding period of ischaemia. Thus, in attempting to assess the ability of any agent to modify reperfusion arrhythmias, the selection of the duration of preceding ischaemia can be critical. For the allopurinol studies we selected a 5 min period of ischaemia (followed by a 10 min period of reperfusion). This choice was based on an earlier characterization study [7]. Using the identical preparation, in which we demonstrated that a 5 min period of ischaemia resulted in the highest incidence of rhythm disturbance, we also showed that all major reperfusion-induced arrhythmias (such as fibrillation and tachycardia) occurred within the first 10 min of reperfusion.

Figure 4 shows that in placebo-treated control animals, the severity of reperfusion-induced arrhythmias tended to be greater than that observed during ischaemia (fig. 3). Thus, almost 70% of animals exhibited ventricular fibrillation during reperfusion, and this lasted much longer than during ischaemia. As a consequence over 50% of rats died, mainly from ventricular fibrillation.

Allopurinol treatment resulted in a major reduction in the incidence of ventricular fibrillation (67–11%; $p < 0.01$) and the mean duration of ventricular fibrillation (230±70 to 14±1 s). Allopurinol also reduced the duration of ventricular tachycardia to approximately one third of that seen in the control group (93±26 to 34±8 s; $p < 0.05$) and halved mortality from 56 to 22%.

Free Radicals and Arrhythmias

It is probably reasonable to state that the development of re-entry circuits (or possibly enhanced automaticity) is the ultimate electrophysiological aberation responsible for the manifestation of arrhythmias (particularly ventricular fibrillation) during ischaemia and reperfusion. It is equally likely that regional heterogeneity of tissue injury (or recovery) is a critical progenitor for the establishment of re-entry circuits. What remains to be established is which of the many complex molecular changes occurring during ischaemia or reperfusion is critical to the initiation of electrophysiological instability. In recent years many processes have been suggested as potential culprits [7, 25, 26]. These include unfavourable redistribution and accumulation of ions (particularly potassium and calcium), the release of catecholamines, changes in cAMP, changes in the availability of glycolytically produced adenosine triphosphate (ATP), the accumulation and utilization of fatty acids, the activity of membrane lysophosphatides and the activity of the α-receptor. To this list of potential arrhythmogenic factors we would add the production of free radicals as a consequence of the activity of the enzyme xanthine oxidase.

In the present study we have shown that the administration of allopurinol prior to coronary artery occlusion and reperfusion in an anaesthetized in vivo rat preparation greatly reduces the incidence and severity of reperfusion-induced arrhythmias and also has some effect upon ischaemia-induced arrhythmias. Association of this observation directly with free radical formation must of course be undertaken with great caution since we have not been able to measure these short-lived molecular intermediates and in this study we have not measured or demonstrated any change in the activity or ratio of xanthine dehydrogenase to xanthine oxidase. Despite this limitation, we would argue that certain other factors support our association between an anti-arrhythmic effect and a reduction of free radical formation. Firstly, allopurinol is a highly specific inhibitor of xanthine oxidase [3]. Secondly, free radical formation will occur under conditions of ischaemia despite the reduced availability of oxygen [36]. However, we would expect there to be far greater production of superoxide during reperfusion because of the return of large quantities of oxygen, this being especially so in very severely ischaemic tissue. This gains further support from our observation that arrhythmias are more severe during reperfusion and that they are more amenable to reduction by allopurinol. Thirdly, in a recent study, *Woodward and Zakaria* [10] reported that the free radical scavenger, superoxide dismutase, can effectively reduce reperfusion arrhythmias in the rat.

Possible Mechanisms for Free Radical Protection. The mechanisms underlying free radical-induced damage in the heart are poorly understood. *Meerson* et al. [37] proposed that superoxide could induce damage by lipid peroxidation, in oxidizing the lipid in membranes a lipid peroxide radical is formed and this would act to break down the structure of the membrane. In addition, the lipid peroxide itself would be a strong oxidant which could initiate further damage. Lipid peroxidation is difficult to detect and at present the only available method has been to assay for malonyl dialdehyde, a by-product of lipid peroxidation. We, as well as others [38, 39], have found that in isolated rat hearts a modest rise in malonyl dialdehyde levels accompanies a 40 min period of global ischaemia or hypoxia.

Recent studies by *Rowe* et al. [42] have suggested that the hydroxyl radical is the real culprit in ischaemic damage. Superoxide can combine with hydrogen peroxide (another product of the xanthine oxidase reaction) and via the Haber-Weiss reaction it can form a hydroxyl radical [40]. It was originally thought that this reaction was too slow to be of pathological significance but it is now known that trace metals in tissue can catalyse the reaction at a much higher rate [41]. Hydroxyl radical scavengers, such as mannitol or catalase, which remove the hydrogen peroxide substrate required for the reaction, have been shown to be able to prevent free radical damage to the sarcoplasmic reticulum [42].

Myocardial tissue contains endogenous scavengers, such as superoxide dismutase, glutathione and catalase, and under normal conditions these provide valuable protection against free radical damage. However, during ischaemia and reperfusion where large quantities of free radicals may be produced these endogenous scavenging systems may be swamped and thus be unable to prevent substantial tissue injury. This problem may be exacerbated by the possibility that in ischaemia these endogenous scavenging systems are themselves inhibited [39].

If free radical formation is able to initiate a sequence of events leading to the development of an arrhythmia, then it is necessary to ask whether this is some direct electrical effect of these charged intermediates or whether they act via some intermediate process. At the present time this question cannot be definitively answered, but we would advance our view that some perturbation of membrane integrity and function may be a key link in the induction of injury and the development of an electrophysiological abnormality. Since it is well-known [3] that free radicals can cause major membrane damage, the most likely effect of free radicals is to alter membrane structure and the membrane's ability to control ionic shifts, particularly that of potassium.

In conclusion, evidence is presented that superoxide radical production from xanthine oxidase may play an important role in the genesis of reperfusion-induced arrhythmias and a lesser role in the genesis of ischaemia-induced arrhythmias and that effective pharmacological control of reperfusion arrhythmias, particularly the potentially lethal rhythm disturbance ventricular fibrillation, might be accomplished either by inhibitors of free radical formation or by the use of appropriate scavengers.

Concluding Comments

A variety of oxygen-derived free radicals are produced as a consequence of normal cellular metabolism. In healthy tissue these potentially toxic metabolic by-products are removed by endogenous scavengers, thus avoiding any serious cellular injury. Under conditions of ischaemia and reperfusion a number of factors may combine to increase the extent of free radical production, initiate new sources of production (e.g. xanthine oxidase) and limit the activity of natural scavenging mechanisms. The resulting increase in free radical activity may well represent an important component in the complex process of tissue injury. It would therefore seem reasonable to include anti-free radical agents in any intervention designed to afford tissue protection.

Evidence has been presented that superoxide and other oxygen-derived free radicals may contribute to the development of myocardial infarction and to the occurrence of arrhythmias. We have also shown that interventions capable of scavenging, or preventing the formation of, free radicals can reduce infarct size and limit the severity of arrhythmias. At this stage the association between free radicals and these events is circumstantial, a problem which arises from the difficulty of measuring these short-lived, highly reactive intermediates. Further studies are required to strengthen this association, to define mechanisms of action and to identify the nature and sources of the free radicals. At the present time, however, we believe that there is significant evidence to warrant further investigations of xanthine oxidase and its ability to produce superoxide.

References

1 Reimer, K. A.; Hill, M. L.; Jennings, R. B.: Prolonged depletion of ATP and of the adenine nucleotide pool due to delayed resynthesis of adenine nucleotides following reversible myocardial ischemic injury in dogs. J. mol. cell. Cardiol. *13:* 229–239 (1981).

2 Bush, L. R.; Buja, L. M.; Samowitz, W.; Rude, R. E.; Walthan, N.; Tilton, G. P.; Willerson, J. T.: Recovery of left ventricular segmental function after long-term reperfusion following temporary coronary occlusion in conscious dogs. Cirulation Res. *53:* 248–263 (1983).

3 McCord, J. M.: Are free radicals a major culprit? in Hearse, Yellon, Therapeutic approaches to myocordial infarct size limitation, pp. 209–218 (Raven Press, New York 1984).

4 Chambers, D. E.; Parks, D. A.; Patterson, G.; Yoshida, S.; Burton, K.; Parmley, L. F.; McCord, J. M; Downey, J. M.: Role of oxygen-derived radicals in myocardial ischemia. Fed. Proc. *42:* 1093 (1983).

5 Parks, D. A.; Bulkley, G. B.; Granger, D. N.; Hamilton, S. R.; McCord, J. M.: Ischemic injury in the cat small intestine. Role of superoxide radicals. Gastroenterology *82:* 9–15 (1977).

6 Hearse, D. J.: Reperfusion of the ischemic myocardium. J. mol. cell. Cardiol. *9:* 605–615 (1977).

7 Manning, A. S.; Hearse, D. J.: Reperfusion-induced arrhythmias: mechanisms and prevention. J. mol. cell. Cardiol. *16:* 497–518 (1984).

8 Hearse, D. J.: Reperfusion of the ischemic myocardium. Clin. Res. Rev. *4:* 58–61 (1984).

9 Mannings, A. S.; Coltart, D. J.; Hearse, D. J.: Ischemia and reperfusion-induced arrhythmias in the rat. Effects of xanthine oxidase inhibition with allopurinol. Circulation Res. *55:* 545–550 (1984).

10 Woodward, B.; Zakaria, M.: The effects of some free radical scavengers on reperfusion-induced arrhythmias in the isolated rat heart. J. mol. cell. Cardiol. *17:* 485–593 (1985).

11 Braunwald, E.; Maroko, P. R.; The reduction of myocardial infarct size. New Engl. J. Med. *291:* 525–526 (1974).

12 Kloner, R. A.; Braunwald, E.: Observations on experimental myocardial ischemia. Cardiovasc. Res. *14:* 371–395 (1980).

13 Opie, L. H.: Myocardial infarct size II. Comparison of anti-infarct effects of beta blockade, glucose-insulin-potassium, nitrates and hyaluronidase. Am. Heart J. *100:* 531–552 (1980).

14 Bache, R. J.: Can drugs really limit infarct size? in Hearse, Yellon, Therapeutic approaches to myocardial infarct size limitation, pp. 185–208 (Raven Press New York 1984).

15 DeWall, R. A.; Vasko, K. A.; Stanley, E. L.; Kezdi, P.: Responses of the myocardium to allopurinol. Am. Heart J. *82:* 362–370 (1971).

16 Akizuki, S.; Yoshida, S.; Chambers, D. E.; Eddy, L.; Parmley, L.; Yellon, D. M.; Downey, J. M.: Blockage of the O_2-radical producing enzyme, xanthine oxidase reduces infarct size in the dog. Fed. Proc. *43:* 1491 (1984).

17 Jolly, S. R.; Kane, W. J.; Bailie, M. B.; Abrams, G. D.; Lucchesi, B. R.: Canine myo-

cardial reperfusion injury; its reduction by the combined administration of superoxide plus superoxide dismutase. Circulation Res. *54:* 277–286 (1984).

18 Romson, J.; Hook, B.; Kunkel, S.; Abrams, G.; Schork, A.; Lucchesi, B.: Reduction in the extent of myocardial ischemic injury by neutrophil depletion in the dog. Circulation *67:* 1016–1023 (1983).

19 Chagrasulis, R. W.; Downey, J. M.: Selective coronary embolization in closed chest dogs. Am. J. Physiol. *233:* 335–337 (1977).

20 Zimmer, H. G.: Effects of inosine on cardiac adenine nucleotide metabolism in rats treated with isoproterenol. J. mol. cell. Cardiol. *15:* suppl., pp. 1–296 (1983).

21 Juhasz-Nagy, A.; Papp, L.; Syabo, Z.: Experimental evidence for the cardioprotective action of inosine. J. mol. cell. Cardiol. *15:* suppl., pp. 1–297 (1983).

22 Krenitoky, T. A.; Tuttle, J. U.; Cattan, E. L.; Weng, P.: A comparison of the distribution and electron acceptor specificities of xanthine oxidase. Compar. Biochem. Physiol. *49:* 687–703 (1974).

23 Schaper, W.: The collateral circulation of the heart (North-Holland, Amsterdam 1971).

23 Lee, J. T.; Idecker, R. E.; Reimer, K. A.: Myocardial infarct size and location in relation to the coronary vascular bed at risk in man. Circulation *64:* 526–631 (1981).

25 Opie, L. H.; Nathan, D.; Lubbe, W. F.: Biochemical aspects of arrhythmogenesis and ventricular fibrillation. Am. J. Cardiol. *43:* 131–148 (1979).

26 Corr, P. B.; Witkowski, F. X.: Potential electrophysiologic mechanisms responsible for dysrhythmias associated with reperfusion of ischemic myocardium. Circulation *68:* suppl. 1, pp. 16–24 (1983).

27 Corr, P. B.; Shayman, J. A.; Kramer, J. B.; Kipnis, R. J.: Increased alpha-adrenergic receptors in ischemic cat myocardium: a potential mediator of electrophysiological derangements. J. clin. Invest. *67:* 1232–1226 (1981).

28 Fitzgerald, J. D.: The effects of beta-adrenoceptor blocking drugs on early arrhythmias in experimental and clinical myocardial ischemia; in Parratt, Early arrhythmias resulting from myocardial ischemia, pp. 295–315 (Macmillan, London 1982).

29 Podzuweit, T.; Dalby, A. J.; Cherry, G. W; Opie, L. H.: Cyclic AMP levels in ischemic and non-ischemic myocardium following coronary arterey ligation: relation to ventricular fibrillation. J. mol. cell. Cardiol. *10:* 81–94 (1978).

30 Corr, P. B.; Gross, R. W.; Sobel, B. E.: Arrhythmogenic amphiphilic lipids and the myocardialcell membrane. J. mol. cell. Cardiol. *14:* 619–626 (1982).

31 Hirche, H.; Friedrich, R.; Kebbell, U.; McDonald, F.; Zykla, V.: Early arrhythmias, myocardial extracellular potassium and pH; in Parratt, Early arrhythmias resulting from myocardial ischemia, pp. 139–154 (Macmillan, London 1982).

32 Clusin, W. T.; Buchbinder, M.; Harrison, D. C.: Calcium overload, 'injury current', and early ischemic cardiac arrhythmias – a direct connection. Lancet 272–273 (1983).

33 Kurien, V. A.; Yates, P. A.; Oliver, M. F.: The role of free fatty acids in the production of ventricular arrhythmias after acute coronary artery occlusion. Eur. J. clin. Invest. *1:* 225–241 (1971).

34 Most, A. S.; Capone, R. J.; Mastrofrancesco, P. A.: Free fatty acids and arrhythmias following coronary artery occlusion in pigs. Cardiovasc. Res. *11:* 198–205 (1977).

35 Selye, H.; Bayusz, E.; Graslo, S.; Mendell, P.: Simple technique for the surgical occlusion of coronary vessels in the rat. Angiology *11:* 395–409 (1960).

36 Rao, P. S.; Cohen, M. V.; Mueller, H. S.: Production of free radicals and lipid peroxides in early experimental myocardial ischemia. J. mol. cell. Cardiol. *15:* 713–716 (1983).

37 Meerson, F. Z.; Kagan, V. E.; Kozlor, Y. P.; Belkina, L. M.; Arkhipinko, Y. P.: The role of lipid peroxidation in pathogenesis of ischemic damage and the anti-oxidant protection of the heart. Basic Res. Cardiol. *77:* 465–485 (1982).

38 Gauduel, Y.; Duvelleroy, M. A.: Role of oxygen radicals in cardiac injury due to reoxygenation. J. mol. cell. Cardiol. *16:* 459–470 (1984).

39 Guarnieri, C.; Flamigni, F.; Caldarera, C. M.: Role of oxygen in the cellular damage induced by reoxygenation of hypoxic heart. J. mol. cell. Cardiol. *12:* 797–808 (1980).

40 Haber, F.; Weiss, J.: The catalytic decomposition of hydrogen peroxide by iron salts. Proc. R. Soc. *A147:* 332–351 (1934).

41 McCord, J. M.; Day, D. E.: The superoxide-dependent production of hydroxyl radical catalyzed by iron-EDTA complex. FEBS Lett. *86:* 139–142 (1978).

42 Rowe, T. G.; Manson, N. H.; Caplan, M.; Hess, M. L.: Hydrogen peroxide and hydroxyl radical mediation of activated leucocyte depression of cardiac sarcoplasmic reticulum. Circulation Res. *53:* 584–591 (1983).

D. J. Hearse, MD, The Rayne Institute, St. Thomas' Hospital,
London SE1 (England)

Novelli, Ursini (eds.), Oxygen Free Radicals in Shock. Int. Workshop,
Florence 1985, pp. 165–168 (Karger, Basel 1986)

Mitochondrial Superoxide Production in Ischemia and Reperfused Rabbit Heart

Carlo Guarnieri, Claudio Muscari, Carlo Ventura,
Claudio M. Caldarera

Istituto Chimica Biologica, Centro Studi e Ricerche sul Metabolismo Cardiaco,
Bologna, Italy

Introduction

Recent studies suggest that the pathogenesis of cardiac ischemic injury [1] and the exacerbation of hypoxic-induced injury by reoxygenation [2] may be ascribed to the production of activated metabolites of oxygen, such as the superoxide radicals (O_2^-) and the OH˙ radicals. These radicals can be formed by soluble enzymes (e.g. xanthine oxidase) or by complex enzymatic systems located in the membrane (microsomes, mitochondria, nuclei) [3]. In the cardiac mitochondria, on the inner membranes, there are two sites of O_2^- production located in the rotenone-inhibited region (complex I) and the antimycin-inhibited ubiquinone-cytochrome b region (complex III) [3]. *Cadenas* et al. [4] showed that the complex I and III supplemented with NADH generated O_2^- at the maximum rates of 9.8 and 6.5 nmol/min mg protein, respectively, while *Turrens and Boveris* [5] indicated that the formation of O_2^- at the level of the NADH dehydrogenase complex is about 50% less than those determined at pH 7.4 at the level of the complex III. Our recent research has revealed that in the heart submitochondrial particles (SMP) supplemented with NADH, the complex I produced 4.01 nmol O_2^-/min mg protein, whilst the complex III 2.20 nmol O_2^-/min mg protein, as determined by the formation of adrenochrome from adrenaline [6]. In the present study, we have investigated if the production of O_2^- by heart SMP changes in consequence of the exposition of the heart muscle to ischemic perfusion followed by reperfusion.

Materials and Methods

The experiments were performed with male New Zealand rabbits. The animals were sacrificed and their hearts perfused by using a Krebs-Henseleit bicarbonate buffer gassed with 95% O_2–5% CO_2 and containing 11 mM glucose as previously described [2]. Ischemia was induced by reducing the coronary flow from 20–25 to 0.5 ml/min for 90 min. The hearts were reperfused for 30 min at the same pre-ischemic value. During all the period of perfusion the hearts were paced at 200 bpm, and maintained at 37 °C. The preparation of mitochondria and SMP were carried out according to previous research [7]. Production of O_2^- was measured spectrophotometrically by following the superoxide dismutase-sensitive oxidation of adrenaline to adrenochrome [8]. The assay system included 250 mM sucrose/50 mM HEPES, pH 7.5, 1 mM adrenaline and 200 µg SMP protein. The production of O_2^- by the complex I was evaluated by incubating heart SMP with 1.5 µM rotenone, while in complex III 2 µM antimycin was used and then the O_2^- rate determined subtracted from that evaluated in the presence of rotenone. Protein determination was made by the method of *Bradford* [9] using bovine serum albumin as standard.

Results

Table I shows that the mitochondrial function evaluated as RCI index is strongly depressed when the heart muscle is exposed to 90 min of ischemic perfusion followed by 30 min of reperfusion. Particularly evident was the negative effect when glutamate, a NAD-linked substrate was used, whilst less evident was the impairment of mitochondrial function, revealed by using succinate, a FAD-linked substrate.

The formation of O_2^- by the complex I enhanced slightly after ischemia (V_m), while the stimulation provoked by the reperfusion was more consistent. The kinetic analysis revealed that in both ischemic and reperfused SMP there was a decreased K_m value for NADH in respect to control SMP. In the complex III for the effect of ischemia and reperfusion, the K_m value did not change, while the V_m value slightly augmented in the reperfused SMP in respect to control.

Discussion

The production of O_2^- radicals in the heart muscle is increased due to the effect of various experimental conditions, such as the incubation of mitochondria with antracyclines [10], or the exposition of rats to hyperoxia [11], or in aged rats [12]. The presence of prolonged cardiac also favors an increase of O_2^- production by heart SMP, particularly at the level of site I [7].

Table I. Effect of ischemia and reperfusion on mitochondrial function (RCI)[a] and on NADH-stimulated O_2^- formation by complex I and III of heart submitochondrial particles

	RCI		Complex I		Complex III	
	glutamate	succinate	K_m[b]	V_m[c]	K_m	V_m
Control	12.5	4.3	0.015	6.5	0.018	3.3
Ischemia	6.8	2.9	0.010	6.7	0.020	3.6
Reperfusion	3.2	2.3	0.010	8.5	0.020	4.1

[a] Respiratory chain index.
[b] mM NADH.
[c] nmol O_2^-/min mg·protein.

The present study shows that SMP prepared from ischemic and re-perfused rabbit heart produces more O_2^- than control SMP, particularly for the effect of reperfusion. This increase is more consistent at the level of the rotenone-inhibited region probably as the consequence of an aug-mented affinity of complex I for the NADH in the ischemic and reper-fused SMP. Since the mitochondrial consumption of NAD-linked sub-strates is depressed in the ischemic and reperfused mitochondria, as re-vealed from the reduced RCI values, it is possible that an accumulation of metabolic NADH in such conditions, maintaining the respiratory chain complex as at a highly reduced state, stimulates the formation of mitochondrial O_2^- radicals.

Less important is such an effect at the level of the antimycin-inhib-ited mitochondrial region, where for the effect of ischemic and reper-fused perfusion, the mitochondrial function judged by the supplementa-tion of succinate is only partially decreased. Therefore, it is highly possi-ble that the NADH-stimulated O_2^- production may be caused by the mi-tochondrial derangements produced by ischemia and reperfusion, partic-ularly at the level of complex I of the respiratory chain.

On the other hand, it is also possible that an increased O_2^- mitochon-drial production, concomitant to a loss of efficiency of the enzymatic O_2^- scavenger mechanisms [13], may be responsible for the formation of mito-chondrial damages in the heart muscle induced by ischemia and espe-cially by reperfusion conditions. Such a hypothesis is supported by ex-periments which, using O_2^- scavengers during ischemic perfusion, im-prove in the cardiac muscle the mitochondrial function [14].

References

1   Rao, P. S.; Cohen, M. V.; Mueller, H. S.: Production of free radicals and lipid peroxides in early experimental myocardial ischemia. J. mol. cell. Cardiol. *15:* 713–716 (1983).

2   Guarnieri, C.; Flamigni, F.; Caldarera, C. M.: Role of oxygen in the cellular damage induced by reoxygenation of hypoxic heart. J. mol. cell. Cardiol. *12:* 797–808 (1980).

3   Freeman, B. A.; Crapo, J. D.: Biology of disease. Free radicals and tissue injury. Lab. Invest. *47:* 412–426 (1982).

4   Cadenas, E.; Boveris, A.; Ragan, C. I.; Stoppani, t. A. O. M.: Production of superoxide radicals and hydrogen peroxide by NADH-ubiquinone reductase and ubiquinol-cytochrome *c* reductase from beef heart mitochondria. Archs Biochem. Biophys. *180:* 248–257 (1977).

5   Turrens, J. F.; Boveris, A.: Generation of superoxide anion by the NADH dehydrogenase of bovine heart mitochondria. Biochem. J. *191:* 421–427 (1980).

6   Guarnieri, C.; Muscari, C.; Ventura, C.; Mavelli, I.: Effect of ischemia on heart submitochondrial superoxide production. Free Rad. Res. Commun. (in press).

7   Guarnieri, C.; Muscari, C.; Caldarera, C. M.: Oxygen radicals and tissue damage in heart hypertrophy; in Harris, Poole-Wilson, Advances in myocardiology, vol. 5, pp. 191–199 (Plenum Publishing, New York 1985).

8   Takeshige, K.; Minakami, S.: NADH- and NADPH-dependent formation of superoxide anions by bovine heart submitochondrial particles and NADH-ubiquinone reductase preparation. Biochem. J. *180:* 129–135 (1979).

9   Bradford, M. M.: A rapid and sensitive method for the quantitation of microgram quantities of protein utilizing the principle of protein-dye binding. Analyt. Biochem. *72:* 246–254 (1976).

10 Davies, K. J. A.; Doroshow, J. H.; Hochstein, P.: Mitochondrial NADH dehydrogenase-catalyzed oxygen radical production by adriamycin and the relative inactivity of 5-iminodannorubicin. FEBS Lett. *153:* 227–230 (1983).

11 Nohl, H.; Heguer, D.; Summer, K.: The mechanism of toxic action of hyperbaric oxygenation and the mitochondria of rat heart cells. Biochem. Pharmacol. *30:* 1753–1757 (1981).

12 Nohl, H.; Hegner, D.: Do mitochondria produce oxygen radicals in vivo? Eur. J. Biochem. *82:* 563–567 (1978).

13 Ferrari, R.; Ceconi, C.; Curello, S.; Guarnieri, C.; Caldarera, C. M.; Albertini, A.; Visioli, O.: Oxygen-mediated myocardial damage during ischemia and reperfusion: role of the cellular defences against oxygen toxicity. J. mol. cell. Cardiol. *17:* 937–945 (1985).

14 Guarnieri, C.; Ferrari, R.; Visioli, O.; Caldarera, C. M.; Nayler, W. G.: Effect of tocopherol on hypoxic perfused and reoxygenated rabbit heart muscle. J. mol. cell. Cardiol. *10:* 893–902 (1978).

Dr. C. Guarnieri, Istituto Chimica Biologica, Centro Studi e Ricerche sul Metabolismo Cardiaco, Via Irnerio 48, I-40126 Bologna (Italy)

Novelli, Ursini (eds.), Oxygen Free Radicals in Shock. Int. Workshop,
Florence 1985, pp. 169–174 (Karger, Basel 1986)

Possible Role of Leukocyte-Derived Oxygen Free Radicals in the Myocardial Failure of Sepsis[1]

*Nancy H. Manson, Mary B. Deardorff, L. Richard Eaton,
Michael L. Hess*

Department of Medicine, Cardiology Division, Medical College of Virginia,
Richmond, Va., USA

Gram-negative sepsis is characterized by the release of a number of chemotaxins which lead to migration, aggregation, and activation of leukocytes [1]. Phagocytosis is accompanied by a respiratory burst which involves the release of oxygen free radicals into the extracellular space. The importance of this respiratory burst is seen in patients with chronic granulomatous disease (CGD) in which leukocytes are normal except that a respiratory burst cannot be initiated. The result is that although bacteria can still be engulfed by the phagocytes, they are not killed in the absence of this respiratory burst [2].

The participation of oxygen free radicals in sepsis has not been thoroughly explored. There have been some elegant studies which support an important role of free radicals in the development of pulmonary consequences of septic shock [3], but the effect of oxygen free radicals on the cardiovascular manifestations of shock has not been fully elucidated.

In order to examine the possible role of oxygen free radicals in myocardial dysfunction, we exposed isolated rabbit right ventricular papillary muscles to leukocytes activated by phorbol myristate acetate and measured mechanical activity of the muscle. Oxygen free radical scavengers were added to the incubation medium in order to identify the species involved in the depression of papillary muscle function.

[1] Supported by HL-30701, a grant-in-aid from the National Institutes of Health.

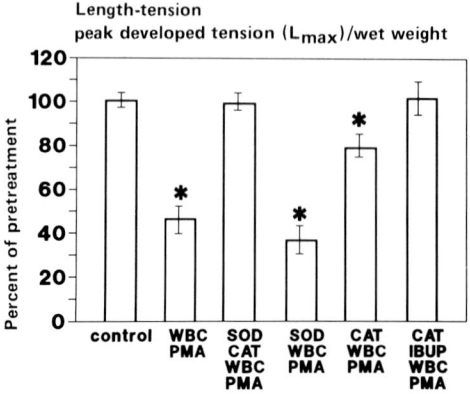

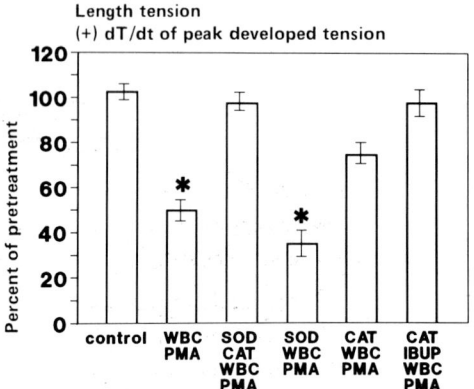

Fig. 1. Length-tension: The effect of phorbol myristate acetate (PMA) activated leukocytes (WBC) alone and in combination with superoxide dismutase (SOD), catalase (CAT), SOD+CAT, or CAT+ibuprofen (IBUP) on peak developed tension and (+) dT/dt both expressed as percent of pretreatment. n=6 for all groups except for CAT+IBUP where n=4. The data are presented as mean ± 1 SEM. *Indicates significant difference (p<0.05) when compared to control.

Methods

Rabbit right papillary muscles were excised, hung vertically in a 5 ml organ bath and perfused with a Krebs-Henseleit bicarbonate buffer (NaCl 137 mM; KCl 4 mM; HCO$_3$ 23.8 mM; H$_2$PO$_4$ 1.8 mM; MgCl$_2$ 0.76 mM; CaCl$_2$ 2.7 mM; D-glucose 5.6 mM) at 37 °C. The buffer was bubbled with 95%O$_2$:5%CO$_2$ to maintain a pH of 7.4. Electrode wires were attached to the muscle and the muscle was attached to a Grass isometric force transducer. Length-tension and force-frequency curves were constructed for each muscle in the presence and absence of treatment. Muscles were divided into

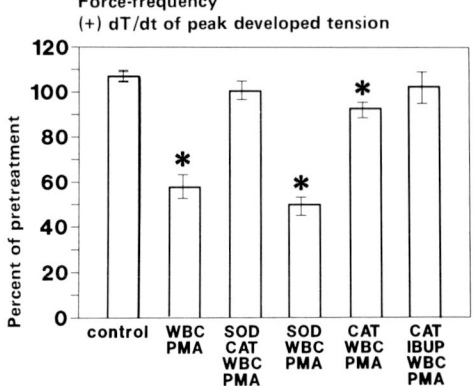

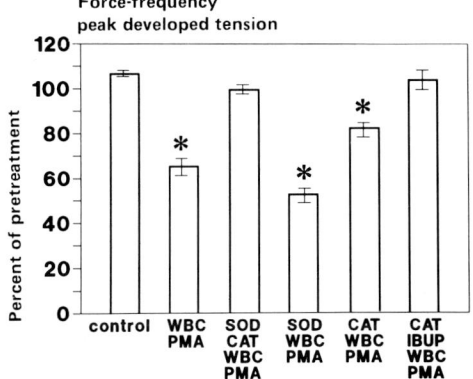

Fig. 2. Force-frequency: The effect of phorbol myristate acetate (PMA) activated leukocytes (WBC) alone and in combination with superoxide dismutase (SOD), catalase (CAT), SOD + CAT, or CAT + ibuprofen (IBUP) on peak developed tension and (+) dT/dt both expressed as percent of pretreatment. n=6 for all groups except for CAT + IBUP where n=4. The data are presented as mean ± 1 SEM. *Indicates significant difference (p < 0.05) when compared to control.

groups with the following agents added to the perfusate: (1) phorbol myristate acetate (PMA; 1 µg/ml) and human leukocytes (8×10⁶ cells/ml); (2) PMA-activated leukocytes (P-W) and superoxide dismutase (SOD; 10 µg/ml); (3) P-W and catalase (CAT; 10 µg/ml); (4) P-W, SOD, and CAT; (5) P-W, CAT and ibuprofen (IBUP; 10 µg/ml). Control groups included muscles exposed to Krebs-Henseleit buffer alone and Krebs-Henseleit and each of the agents listed above alone. Additional studies were performed adding varying concentrations of hydrogen peroxide to the perfusate with or without catalase. Also, baths were run in the presence of xanthine and xanthine oxidase to test the effect of superoxide anion.

Table I. Effects of hydrogen peroxide on peak developed tension (PDT) and dT/dt in rabbit papillary muscle

H_2O_2, mM	Length-tension % pretreatment		Force-frequency % pretreatment	
	PDT, g	dT/dt	PDT, g	dT/dt
0	101.36	104.94	110.79	108.91
0.044	94.08	100.00	90.91	100.00
0.350	82.35	89.74	91.89	93.75
2.000	25.17	28.57	16.00	16.00
4.410	25.97	31.43	18.87	26.32
4.410+CAT (10 µg/ml)	108.57	105.77	111.76	116.67

Table II. Effect of xanthine plus xanthine oxidase on peak developed tension (PDT) and dT/dt in rabbit papillary muscle

Treatment	Length-tension % pretreatment		Force-frequency % pretreatment	
	PDT	dT/dt	PDT	dT/dt
Control	101.36	104.94	110.79	108.91
X+XO	108.76	107.14	111.11	100.00

Results

None of the agents included in the control baths had any effect on length-tension or force-frequency curves. In the length-tension studies (fig. 1), at L_{max}, peak developed tension in the presence of P-W was 49% of control. The first derivative of this tension (dT/dt) was 50% of control. When SOD and CAT or CAT and IBUP were added to the P-W perfusing medium, the muscle was protected (peak developed tension and dT/dt = 100% control). The force-frequency studies (fig. 2) yielded similar results: P-W peak developed tension was 68% of control. DT/dt was 58% of control. When SOD and CAT or CAT and IBUP were added to the perfusing medium, peak developed tension and dT/dt were 100% of control. Table I shows the effect of addition of hydrogen peroxide to the bath. Hydrogen peroxide decreased developed tension and (+) dT/dt obtained from length-tension and force-frequency curves in a concentration-dependent manner.

Table II presents the effect of superoxide anion generation by xanthine-xanthine oxidase (xanthine = 0.1 mM; xanthine oxidase = 0.2 mg/ml) on papillary muscle mechanics. There was no effect on the muscle with this system.

Discussion

Oxygen free radicals are produced by the reduction of oxygen by an NADPH oxidase found in the plasmalemma of phagocytes. It is also possible that some of the free radicals generated during phagocytosis are derived from the arachidonic acid cascade [2]. These free radicals released into the extracellular spaces have been shown to be destructive towards tissues including those of the central nervous system [4], the respiratory system [5], and the cardiovascular system [6].

We conclude from our studies that oxygen free radicals generated by activated leukocytes are negative inotropic agents and that they depress papillary muscle performance. From the results of the studies of the perfusates containing radical scavengers, we conclude that hydrogen peroxide and metabolites of the cyclooxygenase pathway of arachidonic acid breakdown participate in this inhibition of muscle function. IBUP may also be providing a stabilizing effect on the papillary muscle sarcolemma.

References

1   Jacob, H. S.; Craddock, P. R.; Hammerschmidt, D. E.; Moldow, C. F.: Complement-induced granulocyte aggregation: an unsuspected mechanism of disease. New Engl. J. Med. *302:* 789–794 (1980).

2   Fantone, J. C.; Ward, P. A.: Role of oxygen-derived free radicals and metabolites in leukocyte-dependent inflammatory reactions. Am. J. Path. *107:* 397–418 (1982).

3   Shennib, H.; Chiu, R. C.; Mulder, D. S.; Richards, G. K.; Prentis, J.: Pulmonary bacterial clearance and alveolar macrophage function in septic shock lung. Am. Rev. resp. Dis. *130:* 444–449 (1984).

4   Kontos, H. A.; Wei, E. P.; Dietrich, W. D.; Navari, R. M.: Mechanism of cerebral arteriolar abnormalities after acute hypertension. Am. J. Physiol. *240:* H 511–H 527 (1981).

5   Shasby, D. M.; VanBenthuysen, K. M.; Tate, R. M.; Shasby, S. S.; McMurtry, I. F.; Repine, J. E.: Granulocytes mediate acute edematous lung injury in rabbits and

isolated rabbit lungs perfused with phorbol myristate acetate: role of oxygen radicals. Am. Rev. resp. Dis. *125:* 443–447 (1982).

6   Manson, N. H.; Hess, M. L.: Interaction of oxygen free radicals and cardiac sarcoplasmic reticulum: proposed role in the pathogenesis of endotoxin shock. Circulatory Shock *10:* 205–213 (1983).

N.H. Manson, PhD, Box 282, Department of Medicine, Cardiology Division, Medical College of Virginia, Richmond, VA 23298 (USA)

Novelli, Ursini (eds.), Oxygen Free Radicals in Shock. Int. Workshop,
Florence 1985, pp. 175–179 (Karger, Basel 1986)

Chemiluminescence: A Tool To Investigate the Oxidative Stress in the Heart

Renata Barsacchi, Paolo Camici, Gualtiero Pelosi, Nicla Nanni,
Antonio Benassi, Daniela Giannessi, Fulvio Ursini

CNR Institute of Clinical Physiology and Istituto di Patologia Medica I, Pisa, and
Institute of Biological Chemistry, University of Padova, Italy

Introduction

Recent studies suggested a role of lipid peroxidation in the patho-genesis of the ischemic damage of the heart biomembranes [1]. However, a careful evaluation of both free radical generation and lipid peroxida-tion rate in vivo is a difficult task due to the low content and the rapid metabolism of peroxidic compounds to be measured. A useful approach seems to be the chemiluminescence (CL) detection from exposed or per-fused organs, described by *Boveris* et al. [2] and utilized in our laboratory to monitor the oxidative stress in the heart.

The isolated perfused rat heart emits a spontaneous (CL). This CL emission is oxygen-dependent and is stimulated by an oxidative stress in-duced either by the perfusion with micromolar doses of organic hydro-peroxides or by gluthatione-depleting agents [3, 4]. It is generally agreed that CL emission depends on the steady-state hydroperoxil radical con-centration leading to the formation of singlet oxygen or excited carbon-yls. The decay to the ground state of these compounds is accompanied by photon emission that can be recorded on the surface of the organ. More-over, the oxidative stress induces morphofunctional changes resembling those of the so-called 'stone heart' [5], a feature also observed in human pathology which is probably related to a sudden impairment of intracel-lular Ca^{++} homeostasis. The impairment of Ca^{++} homeostasis can be at-tributed to both an ionophoric effect of hydroperoxides [6] and to the in-creased prostaglandins biosynthesis induced by hydroperoxides [7]. Tis-sue hydroperoxides indeed have been reported to regulate prostaglandin synthesis [8].

In the present paper we used a kinetic analysis of the CL emission to evaluate the relationship between the oxidative stress and prostaglandin release from the heart.

Materials and Methods

Hearts were obtained from male Wistar rats (250–300 g body weight) anesthetized with diethyl ether and perfused retrogradely according to Langerdoff. Following a 30-min equilibration with Krebs-Henseleit buffer, gased with 95% O_2 and 5% CO_2, cumene hydroperoxide (CHP) at different concentrations (from 15 to 70 μM was added to the buffer and perfusion continued for a further 20 min.

In a second group of experiments the perfusion medium contained 2.77 mM aspirin since the beginning of the equilibration period; thereafter a 20-min perfusion with CHP was carried out as in the first group of experiments.

CL emitted from the heart was monitored using a single photon-counting apparatus. The terminal portion of the perfusion canula and the heart were enclosed in a light-tight box coupled to an EMI 9658 R photomultiplier, cooled at −11 °C, connected to a minicomputer. The initial emission rate of CL during CHP perfusion, was measured by linear regression analysis of the emission values integrated at 1-second intervals.

The electrocardiogram was continuously monitored by an epicardial electrode. Coronary flow was measured at timed intervals. Timed samples of the perfusate were collected to measure the content of the stable metabolite of PGI_2 (6KetoPGF$_{1\alpha}$) by commercially available (NEN) RIA kit.

At the end of the experiments the hearts were processed for conventional light microscopy using routine histologic stainings. All reagents were of analytical grade.

Results and Discussion

CHP perfusion in the isolated rat heart induced a dose-dependent increase of 6KetoPGF$_{1\alpha}$ release (fig. 1). The release paralleled CL emission (fig. 2). The presence of 2.77 mM aspirin in the perfusate completely inhibited the phenomenon indicating that the increased prostaglandin release is more probably related to the stimulation of the synthesis of prostaglandins than to an aspecific leakage from damaged tissue. In both groups contraction band necrosis increasing at different CHP concentrations from 30 to 80% of the tissue was observed. The histological findings were not significantly modified by aspirin.

The initial rate kinetic analysis of CL emission in the presence of increasing CHP concentrations showed that the phenomenon is sigmoidal

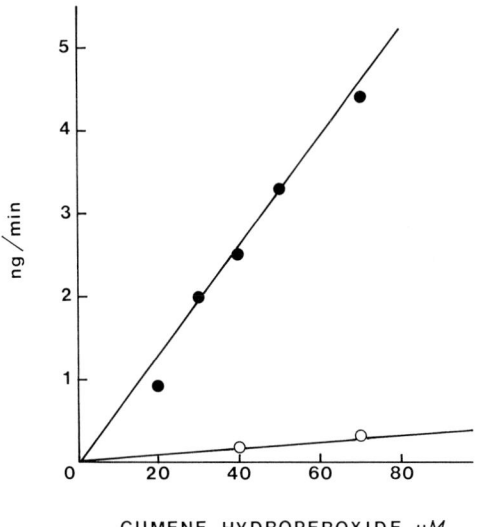

CUMENE HYDROPEROXIDE, μM

Fig. 1. Release of 6KetoPGF$_{1\alpha}$ from perfused rat heart at increasing cumene hydroperoxide concentrations. ○ = Release of 6KetoPGF$_{1\alpha}$ (nmol/min) after 5 min of perfusion with different concentrations of cumene hydroperoxide. ● = Release of 6KetoPGF$_{1\alpha}$ after 5 min of perfusion with different concentrations of cumene hydroperoxide in the presence of 2.77 mM aspirin.

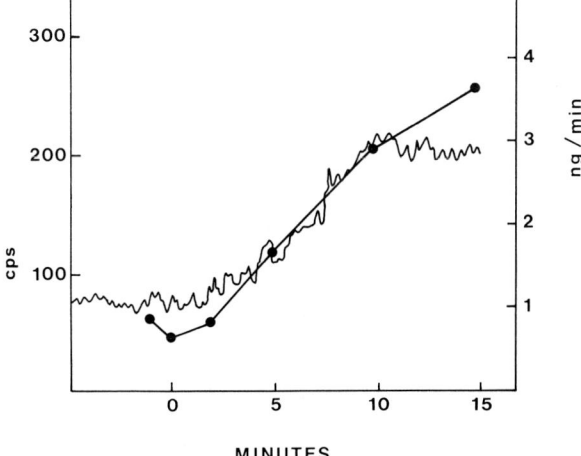

MINUTES

Fig. 2. Chemiluminescence emission and 6KetoPGF$_{1\alpha}$ release (●) from rat heart during perfusion with cumene hydroperoxide 50 μM. For more details see the text.

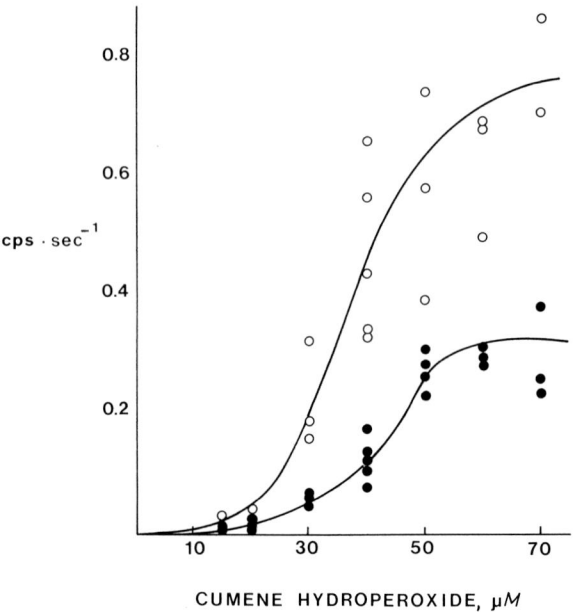

cps · sec^{-1}

CUMENE HYDROPEROXIDE, μM

Fig. 3. Kinetics of chemiluminescence emission induced by cumene hydroperoxide and effect of aspirin. ○ = Initial rate of chemiluminescence increase, during cumene hydroperoxide perfusion. ● = Initial rate of chemiluminescence increase during cumene hydroperoxide perfusion in the presence of 2.77 m*M* aspirin. For more details see the text.

and saturable at 70 μ*M* CHP (fig. 3). Aspirin increased the maximal CL emission rate whilst the saturation was obtained in both cases at the same CHP concentration. Also the CHP concentration at the half of the saturation appeared unmodified by the drug. These results are consistent with a model where the interactions between the hydroperoxide and the membrane structure leading to CL emission are not modified by aspirin. Therefore, the increased CL emission in the presence of aspirin could depend on an increased rate of the reaction leading to light emission. A possible explanation of this phenomenon could be a shift of arachidonic acid cascade induced by hydroperoxide from the cyclooxigenase pathway, inhibited by aspirin, to other, possibly enzymatic, peroxidative pathways.

References

1   Meerson, F. Z.; Kagan, V. E.; Kozlov, Y. P.; Belinka, L. M.; Arkipenko, Y. U.: The role of lipid peroxidation in pathogenesis of ischemic damage and the antioxidant protection of the heart. Basic Res. Cardiol. *77:* 465–485 (1982).

2   Boveris, A.; Cadenas, E.; Reiter, R.; Filipkowski, M.; Nakase, Y.; Chance, B.: Organ chemiluminescence: non-invasive assay for oxidative radicals reactions. Proc. natn. Acad. Sci. USA *77:* 343–351 (1980).

3   Barsacchi, R.; Camici, P.; Bottigli, U.; Salvadori, P. A.; Pelosi, G.; Maiorino, M.; Ursini, F.: Correlation between hydroperoxide-induced chemiluminescence of the heart and its function. Biochim. biophys. Acta *762:* 241–247 (1983).

4   Barsacchi, R.; Pelosi, G.; Camici, P.; Bonaldo, L.; Maiorino, M.; Ursini, F.: Glutathione depletion increases chemiluminescence emission and lipid peroxidation in the heart. Biochim. biophys. Acta *804:* 356–360 (1984).

5   Cooley, D. A.; Reul, G. I.; Wuksch, D. C.: Ischemic contracture of the heart: 'stone heart'. Am. J. Cardiol. *29:* 571–573 (1972).

6   Lebedev, A. V.; Levitsky, D. O.; Loginov, V. A.: Oxygen as inductor of divalent cation permeability through biological and model lipid membranes; in Chrof, Smirnof, Dhalla, Advances in miocardiology, vol. III, pp. 425–438 (Plenum Publishing, New York 1983).

7   Taylor, L.; Menconi, M. J.; Polgar, P.: The participation of hydroperoxides and oxygen radicals in the control of prostaglandins synthesis. J. biol. Chem. *258:* 6855–6857 (1983).

8   Lands, V. E. M.; Kulmacz, R. J.; Marshall, P. J.: Lipid peroxide actions in the regulation of prostaglandins biosynthesis; in Pryor, Free radicals in biology, vol. VI. pp. 39–57 (Academic Press, New York 1984).

R. Barsacchi, MD, CNR Institute of Clinical Physiology, Via Savi 8, I-55100 Pisa (Italy)

Novelli, Ursini (eds.), Oxygen Free Radicals in Shock. Int. Workshop,
Florence 1985, pp. 180–184 (Karger, Basel 1986)

Oxidant-Induced Alterations in Lung Adenine Nucleotides Precede Edema Formation[1]

*H. Redl[a], G. Schlag[a], A. Schiesser[a], S. Bahrami[a], W. Junger[a],
R. G. Spragg[b]*

[a]Ludwig Boltzmann Institute for Experimental Traumatology, Vienna, Austria;
[b]Department of Medicine, UCSD School of Medicine, San Diego, Calif., USA

Introduction

Diffuse acute lung injury in man is thought by most investigators to
result from the complex interaction of numerous biologic mechanisms.
In man, evidence exists which suggests the possibility of injury secondary
to release of oxidants [1, 2] and proteases (e.g. elastase) [1, 3] from neutrophils.

Much useful information concerning mechanisms of acute lung injury has come from studies in intact animal models, in the isolated perfused lung, in cell culture, and in subcellular systems. In particular, studies of lung injury using the isolated perfused rabbit lung (IPRL) have
been useful in studying oxidant injury [4], elastase injury [5], platelet-re-
lated injury [6] and leukotriene D_4-related injury [7]. The isolated lung
has proven a very useful model for establishing the conditions and mediators necessary for the development of acute lung injury. However, the
mechanisms whereby various mediators affect lung injury have not yet
been formally studied.

Although oxidation of proteins, lipids and DNA may play a role in
the cellular and organ dysfunction by oxidant exposure [2, 8, 9], e.g. H_2O_2
from activated neutrophils, specific cellular events resulting in cell death
are yet unknown. *Holmsen and Robkin* [10] found that H_2O_2 caused a
rapid conversion of metabolically available adenosine triphosphate
(ATP) to inosine and/or hypoxanthine in platelets. Endothelial cells

[1] This study was supported by 'Nationalbankfonds'.

have also been noted to undergo alterations in adenylate metabolism after exposure to H_2O_2 or xanthine oxidase/xanthine [11].

Because adenine nucleotides participate in many metabolic reactions and are major factors in maintaining cellular homeostasis, we investigated adenine nucleotide alterations together with permeability changes and release of peroxidation products following exposure of IPRL to H_2O_2 delivered as a bolus.

Because the ratio ATP/adenosine 5'-diphosphate (ADP) and the energy charge (EC) ([ATP + 1/2 ADP]/[ATP + ADP + adenosine monophosphate (AMP)]) of a cell has been related to control of multiple metabolic functions, those values were determined. We have demonstrated a rapid and profound fall in cellular ATP levels and energy charge following exposure to H_2O_2 before the onset of lung edema together with increased levels of thiobarbituric-reactive material (TBA).

Methods

Lung Perfusion

Rabbits weighing 2–3 kg are anesthetized, intubated, heparinized, and lung perfusion is begun in situ. A cell-free balanced salt, 2% albumin, bicarbonate-buffered perfusate is freshly prepared, heated to 37 °C, and the pH is adjusted to 7.40.

Transition from blood to perfusate perfusion is accomplished during less than 15 s. The lungs are perfused at 65 ml/min during which time they are removed from the animal and suspended from a balance in a heated humidified chamber. During this time the perfusate is discarded. A volume of 200 ml of perfusate is then recirculated through the lung at 150 ml/min. Pressure in the perfusion line and the airway is monitored, and ventilation at 10 ml/kg with 2 cm PEEP is continued (fig. 1).

During a 15-min observation period, lung weight, vascular resistance, and airway pressure are monitored. A well-prepared isolated lung (now virtually 100% of experiments) shows no evidence of weight gain or pressure changes over this period of time.

Following administration of H_2O_2 to result in 0, 5 or 30 mM final concentration in perfusion fluid to perfusate over 2 min, observation and recording of lung weight, pulmonary artery and airway pressure are made every 5 min. At the end of the test period, samples of perfusate are collected, lung tissue is sampled by freeze-clamp technique and frozen for subsequent analysis.

Thiobarbituric Acid Reactive Material

One part of lung tissue is homogenized with 10 parts of 0.4 N perchloric acid. 1 ml of the resulting homogenate is mixed with 2 ml TBA reagent (consisting of 0.67% TBA in 2 M Na_2SO_4 according to K. Satoh) and incubated at 100 °C for 15 min. Cold samples are extracted with 3 ml amylalcohol and centrifuged. Supernatant fluid is measured at 530 nm and compared with a standard curve of malondialdehyde (MDA).

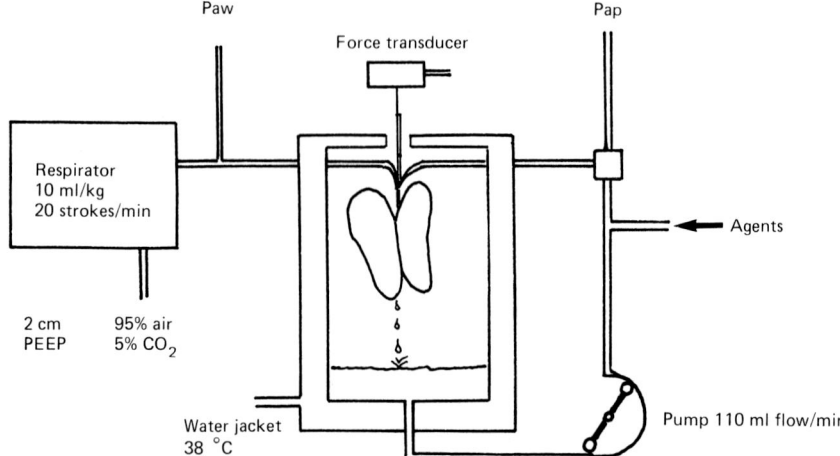

Fig. 1. Schematic diagram of the experimental setup for the isolated perfused rabbit lung. Paw = Airway pressure; Pap = pulmonary artery pressure; PEEP = positive endexpiratory pressure.

Adenine Nucleotides

Frozen tissue is homogenized with 0.4 N HClO$_4$ in a cooled ball mill. The homogenate is neutralized with trioctylamine-Freon [12]. Subsequently nucleotides are separated as reported by *Pogolotti and Sanit* [13].

Results

Isolated perfused rabbit lungs according to this protocol are at least stable for up to 90 min after isolation with practically no weight gain (table I). We have further found adenine nucleotide levels quite similar to the unperfused in situ lungs. Exposure to hydrogen peroxide triggered dose-dependent edema formation with considerable weight gain (up to 26-fold of mean initial total lung weight).

Because of this rather rapid onset of edema, the experiments with 30 mM (group D; tables I, II) had to be terminated within 30 min. Adenine nucleotide levels were markedly decreased (table II). In the group E tissue sampling was done immediately before the dramatic weight gain, but the fall in ATP and the formation of MDA was similar to the group with gross edema (table II).

Table I. Weight gain, pulmonary artery pressure changes and duration (min) of the observation period after H_2O_2 application

Condition H_2O_2	n	Weight gain g	End of perfusion after H_2O_2	Pulmonary artery pressure (–PAP) change mm Hg
B IPRL, 0 mM	3	1.2± 0.3	30.0±0	–
C IPRL, 5 mM	3	17.4±14.7	27.0±3.3	16.0±10.8
D IPRL, 30 mM	3	78.2± 6.8	16.7±7.3	65.7±19.1
E IPRL, 30 mM	3	0.9± 0.3	6.7±1.7	27.0±11.5

IPRL = Isolated perfused rabbit lung.

Table II. Biochemical parameters of IPRL under different experimental conditions

Condition H_2O_2	n	ATP/ADP	EC	Tissue TBA reactive material nmol MDA/g
A In situ	4	11.8±0.4	0.95±0.00	–
B IPRL, 0 mM	3	11.0±1.9	0.94±0.01	86± 23
C IPRL, 5 mM	3	7.2±2.2	0.80±0.08	414±145
D IPRL, 30 mM	3	4.8±0.9	0.70±0.08	–
E IPRL, 30 mM	3	3.6±0.83	0.80±0.01	427±105

Abbreviations are defined in the text.

Conclusion

These results demonstrate that IPRL ATP/ADP and EC are stable for 45 min of perfusion. Significant decreases in ATP/ADP and EC occur in the IPRL following oxidant exposure and are accompanied by evidence of lipid peroxidation. These changes precede the development of overt edema, so one might speculate that an imbalance of the energy-rich phosphate supply due to oxygen radicals results in increased permeability. This might be due to arrest of ATP production as also seen in isolated endothelial cells after peroxide exposure [14] with resultant disturbance of membrane-associated cation ATPase pumps.

References

1 McGuire, W. M.; Spragg, R. G.; Cochrane, C. G.: Studies on the pathogenesis of the adult respiratory distress syndrome. J. clin. Invest. *69:* 543–553 (1982).

2 Cochrane, C. G.; Spragg, S. D.; Revak, S. D.: Studies on the pathogenesis of the adult respiratory distress syndrome: Evidence of oxidant activity in bronchoalveolar lavage fluid. J. Clin. Invest. *71:* 754–761 (1983).

3 Lee, C. T.; Fein, A. M.; Lippman, M.; Holtzman, H.; Kimbel, P.; Weinbaum, G.: Elastolytic activity in pulmonary lavage fluid from patients with adult respiratory distress syndrome. New Engl. J. Med. *304:* 192–196 (1981).

4 Shasby, D. M.; Van Benthuysen, K. M.; Tate, R. M.; Shasby, S. S.; McMurtry, I.; Repine, J. E.: Granulocytes mediate acute edematous lung injury in rabbits and in isolated rabbit lungs perfused with phorbal myristate acetate: role of oxygen radicals. Am. Rev. resp. Dis. *125:* 443–447 (1982).

5 Spragg, R. G.; Lonky, S. A.; Loomis, W. H.; Marsh, J.; Abraham, J. L.: Human neutrophil elastase causes high permeability edema in the isolated perfused rabbit lung (Abstract). Am. Rev. resp. Dis. *125:* suppl., p. 276 (1982).

6 Heffner, J. E.; Tate, R. M.; Canham, E. M.; Shoemaker, S. A.; Patel, M. N.; McMurtry, I. F.; Repine, J. E.: Platelet activating factor stimulates platelets to produce pulmonary hypertension and edematous lung injury in isolated perfused rabbit lungs (Abstract). Am. Rev. resp. Dis. *125:* suppl., p. 277 (1982).

7 Albert, R. K.; Henderson, W. R.: Leukotriene D_4 increases pulmonary vascular permeability in excised rabbit lungs (Abstract). Am. Rev. resp. Dis. *125:* suppl., p. 276 (1982).

8 Kellog, E. W.; Fridovich, I.: Liposome oxidation and erythrocyte lysis by enzymatically generated superoxide and hydrogen peroxide. J. biol. Chem. *752:* 6721–6728 (1977).

9 Brawn, K.; Fridovich, I.: DNA strand scission by enzymatically generated oxygen radicals. Archs Biochem. Biophys. *206:* 414–421 (1981).

10 Holmsen, H.; Robkin, L.: Hydrogen peroxide lowers ATP levels in platelets without altering adenylate energy charge and platelet function. J. biol. Chem. *252:* 1752–1757 (1977).

11 Ager, A.: Assessment of endothelial damage based on responses of cultured aortic endothelial cells to oxygen radicals and hydrogen peroxide; in 1st Int. Endothel. Cell Symp. of the ETCS; The Endothelial Cell – Pluripotent Control Cell of the Vessel Wall, pp. 84–91 (Karger, Basel 1982).

12 Sabina, R. L.; Kernstine, K. H.; Boyd, R. L.; Holmes, E. W.; Swain, J. L.: Metabolism of 5-amino-4-imidazole carboxamide riboside in cardiac and skeletal muscle: effect on purine nucleotides synthesis. J. biol. Chem. *257:* 10178–10183 (1982).

13 Pogolotti, A. L., Jr.; Sanit, D. V.: High pressure liquid chromatography ultraviolet analysis of intracellular nucleotides. Analyt. Biochem. *126:* 335–345 (1982).

14 Spragg, R. G.; Hinshaw, D. B.; Hyslop, P. A.; Schraufstatter, I. U.; Cochrane, C. G.: Alterations in adenosine triphosphate and energy charge in cultured endothelial and $P388D_1$ cells following oxidant injury (submitted for publication).

H. Redl, PhD, Ludwig Boltzmann Institute for Experimental Traumatology, Donaueschingenstrasse 13, A-1200 Wien (Austria)

Novelli, Ursini (eds.), Oxygen Free Radicals in Shock. Int. Workshop,
Florence 1985, pp. 185–188 (Karger, Basel 1986)

Oxygen Free Radicals and Microvascular Injury in a Rabbit Model for ARDS

J. K. S. Nuytinck[a], R. J. A. Goris[a], E. S. Kalter[b], P. H. M. Schillings[c]

Departments of [a] General Surgery, [b] Intensive Care and [c] Pathology,
University Hospital St. Radboud, Nijmegen, The Netherlands

Introduction

We have demonstrated in an earlier study [1] that prolonged activation of the complement system in rabbits, in combination with a short episode of spontaneous hypoxic breathing ($F_iO_2 = 0.1$), results in a picture resembling the adult respiratory distress syndrome (ARDS) in humans. Moreover, in multiple organs we observed accumulation of granulocytes with interstitial and cellular edema. This inflammatory response was interpreted by us as the beginning of multiple organ failure.

Polymorphonuclear neutrophils (PMN), activated by the complement split products C5a and C3a, release lysosomal enzymes, prostanoïds and toxic oxygen radicals that may damage the vascular endothelium, thereby increasing vascular permeability [2, 3].

We have recently tested in vitro rosmarinic acid (RA) (Natterman, Cologne, FRG), a new drug of low toxicity, with complement-inhibiting and anti-oxydative properties and observed no adverse effects on chemotactic, phagocytic and enzymatic killing properties of human PMN [4].

In the present study we administered this drug simultaneously with zymosan-activated plasma (ZAP) to conscious rabbits, both in combination with and without hypoxia, and observed an inhibition of the pulmonary injury, mimicking early ARDS.

Materials and Methods

In unanesthetized New Zealand white rabbits (2.5–3 kg, n = 5), we infused ZAP at a rate of 1 ml/min, during 4 h. Another 5 rabbits received ZAP in combination with a 20-min period of hypoxia ($F_i0_2 = 0.1$) starting 10 min after the onset of the infusion of ZAP. These experiments were repeated while adding intravenous injections of RA 40 mg/kg 10 min before the onset of infusion of ZAP, and 20 mg/kg at intervals of 30 min thereafter.

At selected time intervals, blood was collected from the intravascular lines for blood gas analyses, total and differential white cell count and platelet count. Respiration rate was counted and blood pressure and pulse rate were registrated with a monitor.

4 h after the onset of the infusion with activated plasma, the rabbits were sacrificed by intravenous injection of 4 ml of Nembutal. For histologic studies, the lungs, heart, liver, spleen and right kidney were fixed in formaldehyde 5%, after weighting the heart lung specimen and right kidney. The ratios of heart lung weight/total body weight (HLW/TBW) and right kidney weight/total body weight (RKW/TBW) were calculated. For comparisons within and between groups the sign ranked test and Kruskall-Wallis test were used, respectively. p-values < 0.05 were considered to indicate significant differences.

Results

Hemodynamic Variables. Only small fluctuations were observed in mean arterial blood pressure during the entire protocol and no statistically significant differences were present between the two groups treated with RA and their matching control groups. The pulse rate also remained relatively constant throughout the experiment. Only in the hypoxic control group a significant increase from baseline values was observed between 120 and 150 min after the onset of the infusion.

Respiratory and Blood Gas Variables. In all groups hyperventilation, with a concomitant decrease in P_aCO_2 and increase in pH was observed, especially during the hypoxic episode in the hypoxic groups. No statistically significant differences were observed between the normoxic rabbits receiving RA and their controls with regard to blood gas values and respiratory rate. However, P_aCO_2, bicarbonate and base excess of the hypoxic animals treated with RA decreased significantly less when compared to their hypoxic untreated control animals ($p < 0.01$).

Leukocytes und Platelets. In all groups leukocyte counts decreased significantly from baseline values ($p < 0.01$) and the PMN virtually disappeared from the blood, immediately after the infusion of ZAP. Also, platelets decreased significantly from baseline values in all groups ($p < 0.01$), although less pronounced and less rapidly. In the normoxic rabbits treatment with RA resulted in a significantly lower PMN count after 2 h when compared to their controls. In the hypoxic animals a sig-

Table I. Organ weights (mean ± SD)

Normoxic animals		
HLW/TBW		
Untreated (n = 4)	1.131 ± 0.081	p < 0.01
RA treated (n = 5)	0.917 ± 0.025	
RKW/TBW		
Untreated (n = 4)	0.485 ± 0.032	p < 0.01
RA treated (n = 5)	0.305 ± 0.017	
Hypoxic animals		
HLW/TBW		
Untreated (n = 5)	1.067 ± 0.137	p = 0.02
RA treated (n = 5)	0.894 ± 0.041	
RKW/TBW		
Untreated (n = 5)	0.432 ± 0.039	p < 0.01
RA treated (n = 5)	0.345 ± 0.035	

nificantly higher platelet count was observed during the first 2 h of the experiment, in the group receiving RA.

Pulmonary and Renal Weights. In the normoxic and hypoxic groups receiving RA significantly lower organ weights were found when compared to their matching control groups (table I).

Morphologic Studies. In all groups, equal accumulation of PMN was observed in the lungs. In the normoxic animals equal degrees of modest interstitial edema were observed. However, in the hypoxic animals RA prevented the development of extensive interstitial edema, alveolar edema and hemorrhage observed in the hypoxic control group. The histologic sections of the heart, liver, kidney and spleen revealed that administration of RA resulted in an inhibition of the inflammatory changes, especially in the hypoxic animals.

Discussion

At the onset of the infusion of ZAP in rabbits granulocytopenia developed and at 4 h PMN infiltration was observed throughout multiple organs with increased organ weights. When the rabbits also were subjected to an additional episode of hypoxia (FiO$_2$ = 0.1) frank pulmonary edema formation developed with accentuation of the multiple organ inflammation. RA was able to diminish the increase in organ weights significantly. Also, no clinical picture of ARDS developed in the hypoxic rabbits and hyperventilation and metabolic acidosis were significantly mitigated. Most interestingly, however, RA did not prevent the disappearance of the circulating granulocytes and their adherence and aggregation throughout the capillary networks. It is likely that the generation of oxygen radicals directly or indirectly plays important role in the pathogenic principles of our experimental model. They are produced by the activated granulocytes and are generated after hypoxia during reoxygenation via the xanthine oxidase pathway. Indeed, most pronounced abnormalities were observed in the group infused with ZAP and subjected to a hypoxic episode. Elsewhere in this issue it is described that RA does not affect phagocytosis chemotaxis and oxygen consumption by human granulocytes but decreases cytotoxicity, chemiluminescence, and measurable H_2O_2 production, processes related to external excretion of toxic oxygen species [4]. It is therefore likely that this anti-oxydative mechanism prevails in the actions of RA by which ARDS and multiple organ inflammation are abolished or mitigated in this model.

We conclude that RA is able to prevent pulmonary edema formation and to mitigate signs of multiple organ inflammation in rabbits subjected to an infusion of zymosan-activated homologous plasma and submitted to a hypoxic episode, by virtue of its anti-oxidative properties.

References

1   Nuytinck, J. K. S.; Goris, R. J. A.; Weerts, J. G. E.; Schillings, P. H. M.: Microvascular injury by activated complement: an important mechanism in the pathogenesis of the adult respiratory distress syndrome (ARDS) and multiple organ failure (MOF). Int. J. Microcirc. *3:* 292 (1984).
2   Goldstein, J. M.; Malmsten, C. L.; Samuelsson, B.; Weissman, G.: Prostaglandins, thromboxanes and polymorphonuclear leukocytes: mediation and modulation of inflammation. Inflammation *2:* 309 (1977).
3   Sacks, T.; Moldow, C. F.; Craddock, P. R.; Bowers, T. K.; Jacob, H. S.: Oxygen radicals mediate endothelial damage by complement-stimulated granulocytes. J. clin. Invest. *61:* 1161–1167 (1978).
4   van Kessel, K. P. M.; Kalter, E. S.; Verhoef, J.: Inhibition of oxydative effects of human polymorphonuclear granulocytes by rosmarinic acid (this volume).

J. K. S. Nuytinck, MD, Department of General Surgery, University Hospital
St. Radboud, P. O. Box 9101, NL-6500 HB Nijmegen (The Netherlands)

Novelli, Ursini (eds.), Oxygen Free Radicals in Shock. Int. Workshop,
Florence 1985, pp. 189–192 (Karger, Basel 1986)

An Oxygen Free Radical Demonstrated in Broncho-Alveolar Lavage in Rats Exposed to 100% Oxygen[1]

Daniel Teres[a], Michael Boutin[b], David Ahlberg[b], Alan Dickinson[b]

[a] Critical Care Service, Baystate Medical Center, Springfield, Mass.; Department
of Medicine, Division of Surgery, Tufts University School of Medicine, Boston,
Mass.; [b] Department of Biology and Chemistry, American International College,
Springfield, Mass., USA

Introduction

There is increasing indirect evidence that oxygen metabolites may be
important in the pathophysiology of a variety of acute lung injury states.
For example, in the sheep lung model, a reversible form of acute pulmonary edema can be produced by the injection of either low dose *E. coli* endotoxin or air [1]. In both conditions, there is pulmonary hypertension
and an increase in protein-rich lymphatic flow. Pretreatment with superoxide dismutase prevented pulmonary edema in air embolism but not in
the endotoxin model while pretreatment with catalase reduced lung
lymph flow in both experiments.

Tate et al. [2], using an isolated rabbit lung model, demonstrated that
purine plus xanthine oxidase produced pulmonary edema in the absence
of any cellular elements, presumably through the production of hydrogen
peroxide-derived oxygen metabolites. This is indirect evidence for the
possibly dangerous effects of oxygen metabolites. The objective of the
following study is to attempt to establish direct evidence for the involvement of oxygen free radicals in broncho-alveolar lavage fluid of rats subjected to 97% normobaric oxygen exposure or accelerated oxygen toxicity with antabuse, a substance which interferes with sulfhydryl protective
enzymes [3].

---

[1] Supported by a grant from the Richard and Edith Strauss Canada Foundation.

Materials and Methods

The Production of Oxygen Toxicity. Pathogen free male Sprague-Dawley rats (175–225 grams) were placed in a Plexiglas chamber with >97% atmospheric oxygen for various time intervals. Short-range time intervals included 2, 5, 15, 30 min, 1 and then 2 h. Long-range time intervals included 48, 57, 66, and 72 h. Exposure to the hyperoxic condition in excess of 72 h resulted in the expiration of the animal. Following these specified exposure time intervals, broncho-alveolar lavage was performed on the animals and the fluid tested via electron spin resonance spectrometry (ESR).

Antabuse (Disulfiram). Antabuse is known to accelerate oxygen toxicity in rats and produce severe pulmonary edema following 24h of hyperoxic exposure [3]. 200 mg of antabuse per kilogram of body weight of the rat were injected intraperitoneally prior to placing the rat in the Plexiglas chamber. Exposed intervals to 97% oxygen included 18, 20, 23, and 24 h.

Broncho-Alveolar Lavage Technique. Following hyperoxic exposure, the animals were sacrificed via cervical dislocation of the neck. This was performed in an oxygen atmospheric environment. The trachea was rapidly exposed and lavage fluid injected into the bronchial tubes. The lavage fluid was comprised of the following components: a 10-ml solution with 1 ml of 1 molar spin trap either DMPO (5-dimethyl-1-pyroline-1-oxide) or PBN (1-phenyl-*N*-tertbutyl nitrone) buffered using a pH 7.5 phosphate solution with distilled deionized water. The fluid was then withdrawn from the lungs, rapidly filtered, and then frozen with liquid nitrogen. The filtering was performed to eliminate any cellular components from the lavage solution which might contain antioxidants.

Electron Spin Resonance. Samples were subsequently thawed and placed in the ESR for free radical determination using a Varian E-4 spectrometer. Solutions were placed in a quartz flat cell with peaks and amplitudes recorded during each run. Typically, the settings were as follows: modulation amplitude, 2 Gs (Gauss); microwave power, 100–200 mW; receiver gain, 2000–2500.

Results

No free radicals were present in broncho-alveolar lavage fluid in control animals. For animals maintained in the oxygen chamber for up to 48 h, there were no free radical metabolites present utilizing the technique of thawing the solutions prior to placement in the spectrometer.

At 57 h, broncho-alveolar lavage fluids from rats exposed to >97% normobaric oxygen contained a signal compatible with a carbon based oxygen free radical (fig. 1).

At 66 h, the same signal was present but at a much higher amplitude suggesting increased generation of the carbon-based oxygen free radical.

With antabuse, a faint signal was demonstrated which decayed rapidly after 20 h exposure (fig. 2). At 23 and 24 h after installation of anta-

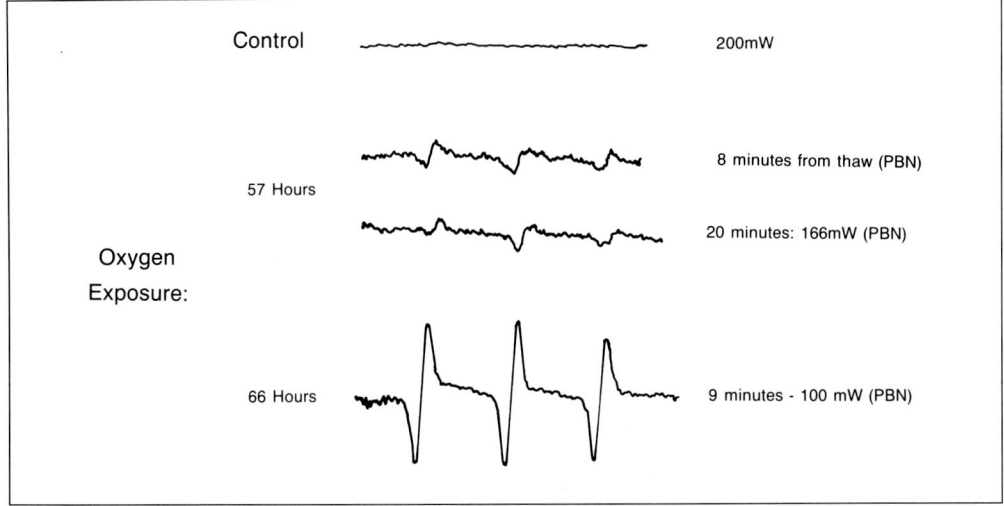

Fig. 1. ESR spectra from broncho-alveolar lavage of animals exposed to oxygen. At 52 and 66 h, a consistent carbon based oxygen free radical is demonstrated.

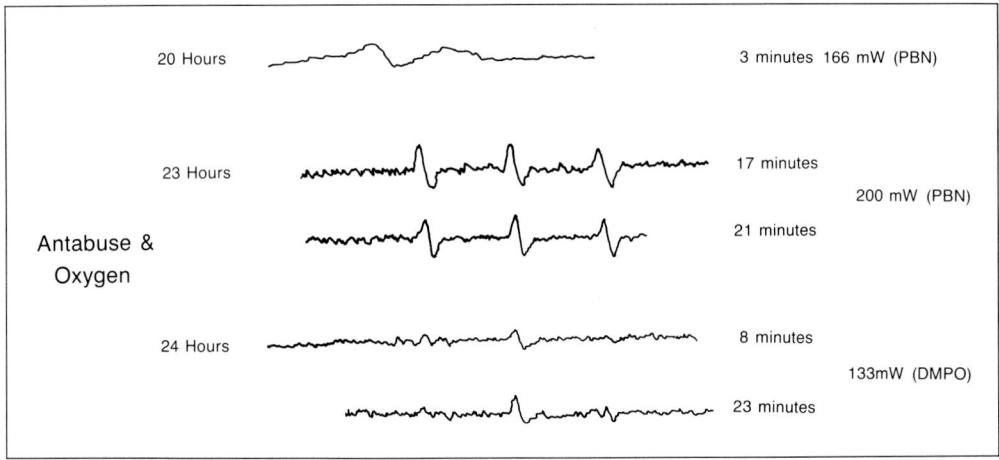

Fig. 2. ESR spectra from broncho-alveolar lavage of rats exposed to antabuse plus oxygen. At 20 h a faint signal is observed. After 23 and 24 h a carbon-based oxygen metabolite is demonstrated.

buse and oxygen exposure, the signal was consistent and stable using both spin traps, DMPO and PBN (fig. 2).

Conclusions

A reactive carbon based oxygen free radical has been demonstrated in a biological model of normobaric oxygen toxicity. Although the exact identification and source of the oxygen free radical is not known, it appears that the signal is present before the large influx of polymorphoneutrophils is expected from other studies [3]. The results suggest that the carbon based radical may be the result of lipid peroxidation of the unit membrane of the endothelial tissues lining the lung.

Future work will be performed by freezing samples into quartz rods that can be fit directly into the ESR spectrometer. This will eliminate the need to thaw specimens since oxygen metabolites are short-lived.

References

1   Flick, M. R.; Hoeffel, J. M.; Staub, N. C.: Superoxide dismutase with heparin prevents increased lung vascular permeability during air emboli in sheep. J. appl. Physiol. *55:* 1284–1291 (1983).
2   Tate, R. M.; Vanbentherysen, K. M.; Shasby, D. M.; McMurtry, I. F.; Repine, J. E.: Oxygen radical mediated permeability edema and vasoconstriction in isolated perfuse rabbit lungs. Am. Rev. resp. Dis. *126:* 802–806 (1982).
3   Deneke, S. M.; Fanburg, B. L.: Normobaric oxygen toxicity of the lung. New Engl. J. Med. *303:* 76–86 (1980).

Daniel Teres, MD, Critical Care Service, Baystate Medical Center,
759 Chestnut Street, Springfield, MA 01199 (USA)

Novelli, Ursini (eds.), Oxygen Free Radicals in Shock. Int. Workshop,
Florence 1985, pp. 193–196 (Karger, Basel 1986)

N-t-Butyl-α-Phenylnitrone Antioxidant Effect on the Alterations in Phospholipid Composition of Alveolar Surfactant and Lungs in Endotoxin Shock Rats

E. Yanev, A. Momchilova, K. Koumanov, G. P. Novelli, N. Nicolov

Department of Pathophysiology, Medical Academy; Central Laboratory of
Biophysics, Bulgarian Academy of Sciences, Sofia, Bulgaria

An important factor for the high mortality rate in gram-negative shock is the disturbance of the lung gas-exchange function with ensuing acute respiratory insufficiency [1,3]. Endotoxin shock gives rise to phospholipid (PL) metabolism disorders [11].

The purpose of our current experiments is to study the alterations of PL in the lungs and alveolar surfactant in endotoxin-treated rats, as well as to assess the effect exerted by the antioxidant N-t-butyl-α-phenylnitrone upon the changes observed.

Material and Methods

8 male Wistar rats with an average body weight of 186 ± 12 g were used in the experiment. The animals were divided into three groups: (1) controls (6 rats); (2) 6 rats treated with *Escherichia coli* (strain 0111: B_4) endotoxin, and (3) 6 rats treated with endotoxin plus antioxidant N-t-butyl-α-phenylnitrone (Sigma, Chemical Company, USA). Endotoxin was injected into the tail vein at dose 13.5 mg/kg body weight (LD_{50}), and antioxidant was injected intraperitoneally at a dose of 150 mg/kg body weight simultaneously with endotoxin. Within 4 h of endotoxin administration, the animals experimented upon were anesthetized with nembutal 30 mg/kg body weight, and then tracheotomized. The trachea was cannulated and bronchoalveolar lavage was obtained by intratracheal washing with four portions of saline up to a final volume of 20 ml [5]. The lungs of all animals were collected for biochemical study. PL were extracted by the method of *Kahovkova and Odavic* [7], and were determined by the inorganic phosphorus according to the method of *Folch* et al. [6] by thin-layer chromatography. Protein assessment was done after the method of *Lowry* et al. [8]. The results were processed for statistical significance using Student's t-test.

Table I. Phospholipid composition of alveolar surfactant from rats (μgPL/mgPr; mean ± SD)

	Control	Endotoxin	Endotoxin + antioxidant
Sphingomyelin	60.35± 1.07	52.72± 3.14*	47.17± 2.17
Phosphatidylcholine	2,041.01±40.64	1,306.40±20.29*	1,704.14±38.27*
Phosphatidylserine	58.16± 1.12	59.74± 2.03	49.15± 1.16
Phosphatidylinositol	60.32± 0.89	57.14± 2.17	58.06± 1.35
Phosphatidylethanolamine	114.08± 1.70	98.08± 2.15	106.39± 1.02*
Phosphatidylglycerol	340.14± 4.23	288.12± 7.81*	296.12± 4.12
Total phospholipids	2,673.95	1,864.21	2,261.03

* $p < 0.05$.

Results and Discussion

Table I presents the PL composition of alveolar surfactant in the three groups of rats. It is evident that in endotoxin-treated rats the two most important surface-active PL – phosphatidylcholine (PC) and phosphatidylglycerol (PG) – were significantly decreased ($p < 0.05$). It is of interest to note that in rats treated with endotoxin and antioxidant PC level was increased as compared to rats treated with endotoxin alone. Furthermore, PG was reduced in endotoxin-treated rats, but it was very slightly augmented in rats treated with both endotoxin and antioxidant. The rest of the PL fractions which are not essential for the proper functioning of alveolar surfactant were changed as follows: sphingomyelin (SM) was diminished in endotoxin-treated rats, and was not restored in rats treated with endotoxin and antioxidant. Phosphatidylserine (PS) and phosphatidylinositol (PI) remained practically unchanged, while the level of PG was reduced in endotoxin-treated rats, and was very slightly and statistically insignificantly elevated in the animals of group three.

The lung PL composition of all three animal groups disclosed slight and insignificant alterations which are presented in table II. It seems obvious that the changes in PL composition of the alveolar surfactant are much more significant than those in the lung PL.

The results obtained in rats with endotoxin shock may be explained by endotoxin activation of the oxygen free radicals formation [9] which in turn provoke secondary peroxidation of the unsaturated fatty acids in the

Table II. Phospholipid composition of lung tissue from rats (μgPL/g tissue; mean ± SD)

	Control	Endotoxin	Endotoxin + antioxidant
Sphingomyelin	1.670±0.38	1.460±0.67	1.480±0.51
Phosphatidylcholine	8.980±0.57	8.160±0.46	8.280±0.24
Phosphatidylserine	1.640±0.41	1.270±0.21	1.310±0.22
Phosphatidylinositol	1.012±0.38	1.162±0.11	1.206±0.31
Phosphatidylethanolamine	3.740±0.39	3.680±0.14	3.669±0.18
Phosphatidylglycerol	1.480±0.24	1.507±0.20	1.540±0.27
Diphosphatidylglycerol	0.180±0.06	0.174±0.09	0.175±0.04
Total phospholipids	18.702	17.404	17.659

cell membrane PL [4,10], and of the alveolar surfactant as well [2]. The pathogenetic mechanism outlined is suggested on the grounds of partial normalization of the PL composition of alveolar surfactant following antioxidant treatment.

References

1 Birkun, A. A.; Nesterov, E. N.; Kobozev, G. V.: Lung surfactant (Zdorovja, Kiev, 1981).

2 Kozlov, I. A.; Vaiginina, M. A.; Meshteriakov, G. A.: Lung surfactant system. Anesth. Analg. Réanim. 2; 68–71 (1984).

3 Clowes, G. H. A.: Pulmonary abnormalities in sepsis. Surg. Clinss. N. Am. 54: 993–1013 (1974).

4 Del Maestro, R. F.: An approach to free radicals in medicine and biology. Acta physiol. scand. 492: 153–168 (1980).

5 Etoh, T.; Kakishita, E.; Nagai, K.: Role of alveolar macrophage plasminogen activator in the acute pulmonary responses to endotoxin. Lung 162: 49–58 (1984).

6 Folch, J.; Lees, M.; Sloane-Stanley, G. H.: A simple method for the isolation and purification of total lipids from animal tissues. J. biol. Chem. 226; 497–509 (1947).

7 Kahovkova, J.; Odavic, R.: A simple method for analysis of phospholipids separated by thin layer chromatography. J. Chromatogr., biomed. Appl. 40: 90–95 (1969).

8 Lowry, H. J.; Rosenbrough, N. J.; Far, L.; Randay, R.: Protein measurement with the Folin reagent. J. biol. Chem. 193: 265–269 (1951).

9 Noelli, G. P.; De Gaudo, A. P.: Oxygen-free radicals in shock states; in Lewis, Haglund. Shock research, 31–40 (Elsevier, Amsterdam 1983).

10 Slater, T. F.: Free-radical mechanisms in tissue injury. Biochem. J. *222:* 1–15 (984).
11 Von Wichert, P.; Temmesfeld, M.; Meyer, W.: Influence of septic shock upon phosphatidylcholine remodelling mechanism in rat lung. Biochim. biophys. Acta *664:* 487–497 (1981).

Prof. N. A. Nicolov, MD, Department of Pathophysiology, 1, G. Sofiisky Street, Sofia 1431 (Bulgaria)

Novelli, Ursini (eds.), Oxygen Free Radicals in Shock. Int. Workshop, Florence 1985, pp.197–204 (Karger, Basel 1986)

Oxyradicals and Acute Gastrointestinal Mucosal Damage[1]

U. Haglund, S. Arvidsson, M. H. Schoenberg

Departments of Surgery, University of Lund; Malmö General Hospital, Malmö, Sweden; University of Lübeck, FRG

Acute gastrointestinal mucosal lesions may develop following severe injury, sepsis and shock. These complications constitute parts of the multiple organ failure syndrome [6, 11, 13]. The pathogenesis of the development of acute gastrointestinal mucosal damage, however, remains uncertain. Below are summarized experiments aimed to explore the possible role of oxygen-derived free radicals in the development of acute gastrointestinal mucosal ulcerations in cats subjected to local intestinal ischemia or to live *Escherichia coli* bacteremic shock. Detailed descriptions of these experiments have been given elsewhere [3, 25].

Material and Methods

The experiments were performed on cats anesthetized with pentobarbital (Pentothal; 50 mg) and chloralose (50 mg/kg) or with ketamine-HCl (Ketanest: 10 mg/kg) and xylazine (Rompur; 2.5 mg/kg). The cats were tracheotomized and ventilated artificially. They were all given a slow i.v. infusion of 10% glucose containing 10 mmol $NaHCO_3$/100 ml (6 ml/h).

Local Intestinal Ischemia
After laparotomy the duodenum, the spleen, the omentum and the colon were extirpated. A segment corresponding to 75% of the small intestine was isolated with intact vascular supply from the superior mesenteric artery. The splanchnic nerves were cut and the distal ends put on a platinum electrode for stimulation during ischemia. By means of an adjustable clamp around the superior mesenteric artery the intestinal arte-

[1] This study was sponsored by the Swedish Medical Research Council (project no. 4502) and Deutsche Forschungsgemeinschaft (SCH 309:1).

rial blood pressure was reduced to 25–30 mm Hg. After 2 h the clamp was released and the cats were observed additionally for 1 h. During the prehypotensive period, just before, 10 and 60 min after the release of the clamp, intestinal tissue samples were excised for histological examination. In total 21 cats were studied. 7 of these, chosen randomly, received, in addition, at 60 min of hypotension an i.v. injection of 15,000 U/kg yeast superoxide dismutase (SOD) purified according to *McCord and Fridovich* [19]. Since SOD is excreted rapidly by the kidneys [15] the renal arteries and veins were ligated bilaterally prior to the i.v. injection.

Live E. coli *Bacteremia*

After laparotomy and splenectomy the hepatic artery was ligated proximally to the gastrodudenal artery; allowing collateral circulation to the liver from the superior mesenteric artery. An electromagnetic flow-probe (Cliniflow, model 601 D) was placed on the celiac artery. The intragastric pressure was kept at 2 cm of water and bile from the gallbladder (0.6 ml/kg) and 80 mM HCl was instilled into the gastric lumen through a duodenogastric tube.

The cats were thoracotomized and a catheter was placed in the left atrium of the heart to allow determinations of blood flow distribution to the tissues by means of radioactively labeled microspheres [for details see ref. 3, 4]. The principles of this technique, as described by *Heymann* et al. [14], were adhered to. Microspheres were injected before bacteria, at 15 and approximately 150 min after the start of the bacterial infusion. Specimens were taken from the mid-part of the small intestine and from three predetermined areas from the gastric fundus.

Bacteremia was induced by i.v. infusion of an *E. coli* strain (*E. coli* 06K 13 H1) (WHO designation SU 4344/41) from the WHO Collaborative Centre for Reference and Research on *Escherichia* (State Serum Institute, Copenhagen, Denmark). Just prior to the infusion the bacteria were washed in saline, resuspended in 0.9% NaCl, and infused at a concentration of 10^9 live cells/ml. All cats received 1 ml/(kg × min) for 2 min and then 1 ml/(kg × h) for up to 3 h. One series of animals (n=8) also received yeast CuZn superoxide dismutase (SOD) (Pharmacia, Uppsala, Sweden; 5 mg as an i.v. bolus 2 min before bacterial infusion followed by a continuous i.v. infusion of 50 mg/3 h). In addition, 4 of these cats were also given bovine catalase (Boehringer-Ingelheim AB, Skärholmen, Stockholm, Sweden) in the same dose. The renal vessels were not ligated in these 8 animals. Another series of animals (n=8) had no SOD/catalase treatment and served as controls.

Histological Examination

The specimens were prepared by routine, stained with hematoxylin-eosin and coded. The small intestinal mucosa was graded from 0 to 5 according to *Chiu* et al. [7], adapted to cats [1, 8]. Grade 0 means normal mucosa and grade 1 development of a subepithelial space at the tip of the villus. Grade 2 is characterized by epithelial lifting and in grade 3 the lifting is more massive, including also the sides of the villus. In grade 4 the villus is denuded and in grade 5 there is disintegration of the mucosal lamina propria, hemorrhage and ulceration.

The gastric mucosal lesions were graded 0–4 [2]. Grade 0 means normal mucosa. Grade 1 means edema beneath superficial epithelium. In these two grades the superficial epithelium is intact. In grade 2 the surface epithelial cells have disappeared. The upper half of the glands are damaged in grade 3. In grade 4 the glands have disappeared. The gastric mucosal damage was assessed by gastric index (calculated by adding the grade obtained in the three different areas in each animal) or by the number of areas with intact superficial epithelium (grade 0–1) in the different series.

Results

Almost all small intestinal specimens obtained before the period of regional intestinal hypotension disclosed a normal mucosa. At the end of the regional hypotensive period most cats had developed a moderate but still significant mucosal damage (grade 2–3; fig. 1). 10 min after the reperfusion, and even more so 60 min after the reperfusion, the mucosal damage was statistically significantly aggravated. 60 min after the reperfusion the villi were completely denuded and frequently hemorrhagic ulcerations were seen, i.e. grade 4–5. SOD treatment did not influence the development of small intestinal mucosal damage during the hypotension. However, the aggravation during the reperfusion was prevented by SOD and the degree of small intestinal mucosal damage at the end of the hypotensive phase and 60 min after reperfusion were closely similar.

In the bacteremic shock experiments intestinal mucosal tissue was only obtained at the end of the experiments. Earlier experiments in our laboratory have, however, demonstrated that also following this preparation the gastric and the small intestinal mucosa are normal and remain so in sham shock animals [2, 4]. Untreated bacteremic cats had a median gastric mucosal index of 8 with an interquartile range of 7–8. Corresponding values for SOD/catalase-pretreated cats were 5 and 3–7 (p < 0.1 vs controls). Comparing the total number of areas examined (three predetermined areas in each cat) significantly more areas with pronounced lesions (grade 2–4) were found in control cats compared to SOD/catalase-treated (p < 0.002). In figure 2 the gastric mucosal damage is depicted area by area. As demonstrated in figure 2, there is a statistically significant difference only in the major curvature, although there is a tendency to difference also in the anterior and the posterior wall.

In the untreated control animals 4 of the 8 cats had a normal small intestinal mucosa at the end of the bacteremic period. 3 cats had mucosal damage graded 2–3 and 1 cat 4–5. Of the SOD/catalase-pretreated cats 5 had grade 0–1 damage and 3 grade 2–3. There was, thus, no clear-cut difference between the two series of animals as regards the small intestinal mucosal damage.

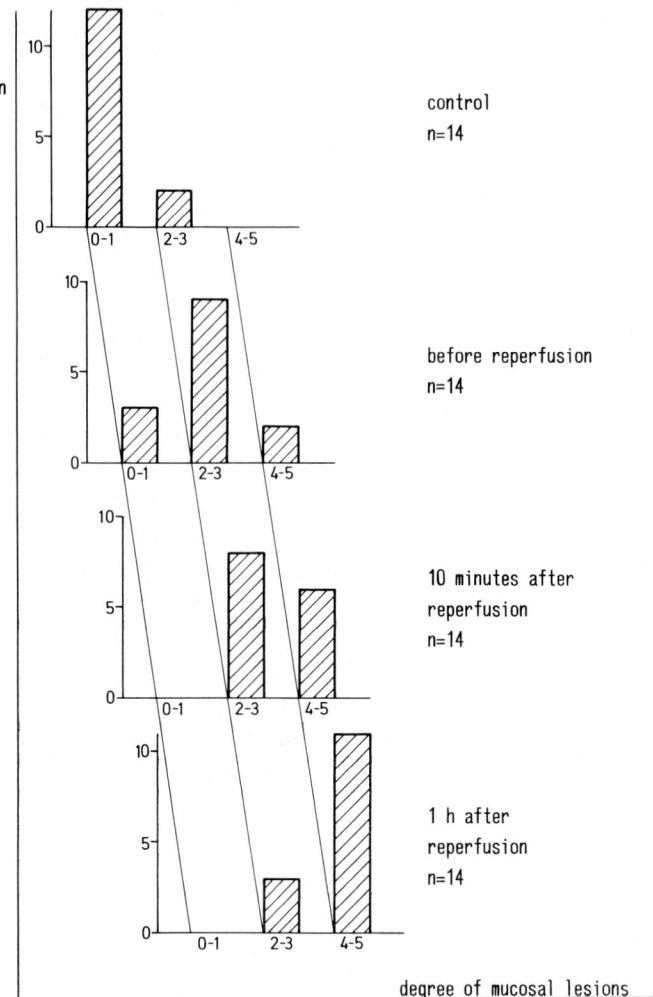

Fig. 1. Grade of small intestinal mucosal damage before, at the end and 10 and 60 min after a 2-hour period of regional intestinal hypotension. From *Schoenberg* et al. [25] with permission.

Discussion

The present series of experiments demonstrate that small intestinal mucosal damage develops during ischemia and during septicemia. There was no evidence obtained that this development was influenced by SOD

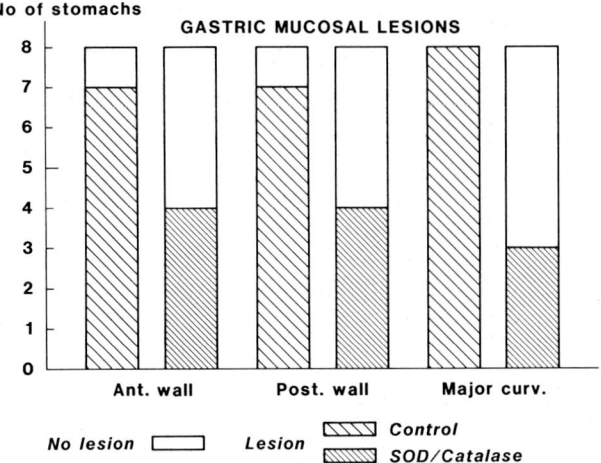

Fig. 2. Number of stomachs with normal or damaged mucosa (grade 0–1 or 2–4, respectively) in specimens obtained from the fundic anterior and posterior wall and from the fundic part of the greater curvature. Data obtained from 8 cats given live *E. coli* i. v. [see ref. 2 for details].

or SOD/catalase, enzymes scavenging oxygen-derived free radicals. In the reperfusion phase following local ischemia there was a significant aggravation of the small intestinal mucosal damage. This aggravation was prevented by SOD indicating that it was caused by generated oxygen-derived free radicals. The acute gastric mucosal damage during septicemia was attenuated but not prevented by SOD/catalase indicating a role of free radicals in the pathogenesis of these lesions, although the mechanism here seems to be more complex.

It was expected from previous experiments that about 50% of the bacteremic cats should develop small intestinal mucosal damage. Furthermore, it has been demonstrated that the development of small intestinal mucosal damage in septic shock is correlated to hypotension [8] but not to mucosal blood flow [9]. Also in regional intestinal ischemia the mucosal blood flow remains rather normal [23]. On the other hand, it has been demonstrated that the mucosal damage following ischemia and septicemia is caused by hypoxia [1, 9]. This paradox could be explained by the villus countercurrent exchange mechanism as discussed in detail elsewhere [8, 18].

The pathophysiological mechanism behind acute gastric ulcerations is controversial. Acid and bile, two factors which were supplied exogenously in the present experimental model, are generally believed to be important [17, 24]. Mucosal ischemia has also been proposed as one important factor [17] but recent experiments indicate that ischemia is not important for the development of acute gastric ulcerations in septicemia and septic shock [2, 12, 22].

Intracellular enzymes can induce the production of superoxide radicals [10, 20]. One such enzyme, which is particularly abundant in the intestines, is the xanthine dehydrogenase/oxidase system. This enzyme catalyzes the oxidation of hypoxanthine to xanthine. If catalyzed by xanthine oxidase, which is the case following ischemia, superoxide radicals are generated [21, 25]. The importance of this mechanism for development of mucosal damage was supported in recent experiments using the competitive inhibitor allopurinol [26]. In addition, polymorphonuclear leukocytes are known to be activated in septicemia, and the respiratory burst of phagocytosing leukocytes leads to a release of free radicals [5, 16]. Whichever the mechanism of generation, oxygen-derived free radicals may result in peroxidative injury of membranes of cells, organelles and of DNA.

To summarize, the present data indicate an important role of oxygen-derived free radicals in the pathogenesis of the reperfusion damage to the small intestine and in the pathogenesis of acute gastric mucosal ulcerations in septicemia but not in the pathogenesis of acute mucosal damage of the small intestine in bacteremic shock.

References

1 Åhrén, C.; Haglund, U.: Mucosal lesions in the small intestine of the cat during low flow. Acta physiol. scand. *88:* 541–550 (1973).

2 Arvidsson, S.; Fält, K.; Haglund, U.: Acute gastric mucosal ulcerations in septic shock. An experimental study on pathogenic mechanisms. Acta chir. scand. *150:* 541–547 (1984).

3 Arvidsson, S.; Fält, K.; Marklund, S.; Haglund, U.: Role of free oxygen radicals in the development of gastrointestinal mucosal damage in *Escherichia coli* sepsis. Circul. Shock *16:* 383–393 (1985).

4 Arvidsson, S.; Lindblad, B.; Esquivel, C.; Fält, K.; Lindström, C.; Bergqvist, D.; Haglund, U.: The effects of dihydroergotamine on the feline cardiovascular response to i. v. infusion of live *E. coli* bacteria. Eur. surg. Res. *16:* 220–231 (1984).

5 Babior, B. M.: Oxygen-dependent microbial killing by phagocytes. New Engl. J. Med. *298:* 659–668 (1978).

6 Borzotta, A. P.; Polk, H. C.: Multiple system organ failure. Surg. Clins Am. *63:* 315–336 (1983).

7 Chiu, C.-J.; McArdle, A. H.; Brown, R.; Scott, H. J.; Gurd, F. N.: Intestinal muco-sal lesion in low-flow states. Archs Surg., Chicago *101:* 478–483 (1970).

8 Falk, A.; Myrvold, H. E.; Lundgren, O.; Haglund, U.: Mucosal lesions in the feline small intestine in septic shock. Circul. Shock *9:* 27–35 (1982).

9 Falk, A.; Redfors, S.; Myrvold, H. E.; Haglund, U.: Small intestinal mucosal le-sions in feline septic shock – a study on the pathogenesis. Circul. Shock (in press).

10 Freeman, B. A.; Crapo, J. D.: Biology of disease. Free radicals and tissue injury. Lab. Invest. *47:* 412–426 (1982).

11 Fry, D. E.; Pearlstein, L.; Fulton, R. L.; Polk, H. C.: Multiple system organ failure. The role of uncontrolled infection. Archs Surg., Chicago *115:* 136–140 (1980).

12 Genter, B.; Stone, A. M.; Stein, T. A.; Wise, L.: Gastric mucosal blood flow and *Es-cherichia coli* bacteriema. Am. J. Surg. *145:* 364–368 (1983).

13 Haglund, U.: Gastrointestinal, hepatic and renal complications of shock and trauma; in Little, Frayn, The scientific basis of the care of the critically ill (Man-chester University Press, Manchester 1985).

14 Heymann, M. A.; Payne, B. D.; Hoffman, J. I. E.; Rudolph, A. M.: Blood flow measurements with radionuclide-labeled particles. Prog. cardiovasc. Dis. *20:* 55–79 (1977).

15 Huber, W.; Saifer, M. G. P.: Orgotein, a drug version of bovine Cu Zn superoxide dismutase. A summary account of safety and pharmacology in laboratory ani-mals; in Michelson, McCord, Fridovich, Superoxide and superoxide dismutase, p. 517 (Academic Press, New York 1977).

16 Jacob, H. S.: Granulocyte-complement interaction. A beneficial antimicrobial mechanism that can cause disease. Archs intern. Med. *138:* 461–463 (1978).

17 Kivilaakso, E.; Silen, W.: Pathogenesis of experimental gastric-mucosal injury. New Engl. J. Med. *301:* 364–369 (1979).

18 Lundgren, O.; Haglund, U.: The pathophysiology of the intestinal countercurrent exchanger. Life Sci. *23:* 1411–1422 (1978).

19 McCord, J. M.; Fridovich, I.: Superoxide dismutase: an enzymatic function of erythrocuprein. J. biol. Can. *244:* 6049 (1969).

20 McCord, J. M.; Fridovich, I.: The biology and pathology of oxygen radicals. Ann. intern. Med. *89:* 122–127 (1978).

21 Parks, D. A.; Bulkley, G. B.; Granger, D. N.; Hamilton, S. R.; McCord, J. M.: Is-chemic injury in the cat small intestine: role of superoxide radicals. Gastroenter-ology *82:* 9–15 (1982).

22 Payne, J. G.; Bowen, J. C.: Hypoxia of canine gastric mucosa caused by *Escheri-chia coli* sepsis and prevented with methylprednisolone therapy. Gastroenterology *80:* 84–93 (1981).

23 Redfors, S.; Hallbäck, D. A.; Haglund, U.; Jodal, M.; Lundgren, O.: Blood flow distribution, villous osmolality and fluid and electrolyte transport in the cat small intestine during regional hypotension. Acta physiol. scand. *121:* 193–209 (1984).

24 Rees, M.; Bowen, J. C.: Stress ulcers during live *Escherichia coli* sepsis. The role of acid and bile. Ann. Surg. *195:* 646–651 (1982).

25 Schoenberg, M. H.; Muhl, E.; Sellin, D.; Younes, M.; Schildberg, F. W.; Haglund, U.: Posthypotensive generation of superoxide free radicals – possible role in the pathogenesis of the intestinal mucosal damage. Acta chir. scand. *150:* 301–309 (1984).

26 Schoenberg, M. H.; Fredholm, B.; Haglund, U.; Jung, H.; Sellin, D.; Younes, M.;
    Schildberg, F. W.: Studies on the oxygen radical mechanism involved in the small
    intestinal reperfusion damage. Acta physiol. scand. *124:* 581–589 (1985).

U. Haglund, MD, Department of Surgery, Malmö General Hospital,
S-214 01 Malmö (Sweden)

Novelli, Ursini (eds.), Oxygen Free Radicals in Shock. Int. Workshop,
Florence 1985, pp. 205–211 (Karger, Basel 1986)

Protection of Liver Allografts from Ischemic Damage Prior to Transplantation Using Insulin and Catalase

Luis H. Toledo-Pereyra, John Cederna

Department of Surgery, Section of Surgical Research, Mount Carmel Mercy
Hospital, Detroit, Mich., USA

Introduction

The deleterious effects of warm ischemia on the subsequent function
of hepatic allografts has been partially traced to the production of oxygen-derived free radicals such as OH^- and O_2^- [1]. Recent work in our
laboratories (also published in this volume), has indicated that pretreatment of hepatic grafts with catalase, an oxygen free radical scavenger,
could improve the function of ischemically damaged canine livers tested
on an isolated perfusion system. In addition, insulin has been identified
as an important hepatotrophic factor [2, 3]. The purpose of the present
study was to assess whether pretreatment of liver grafts with catalase
and/or insulin prior to ischemic damage would improve function after
subsequent transplantation.

Materials and Methods

This study was designed to test the effects of various pretreatment modalities,
given via the portal vein, 5 min prior to iatrogenic ischemic damage. Ischemia was induced by cross-clamping of the hepatic artery and portal vein for 20 min (fig. 1). Donor
animals were divided into the following groups: group I (n = 12) control, no pretreatment, group II (n = 16) lente insulin (1,000 U/rat), group III (n = 10) lente insulin (100
U/rat), group IV (n = 10) catalase (5,000 U/kg), and group V (n = 10) catalase (5,000
U/kg) + lente insulin (1,000 U/rat).

Adult female CD rats (Charles River), weighing between 275 and 300 g were
anesthetized using nembutal (0.5 ml/rat). Through a midline incision, the hepatic ar-

Donor Liver Pretreatment

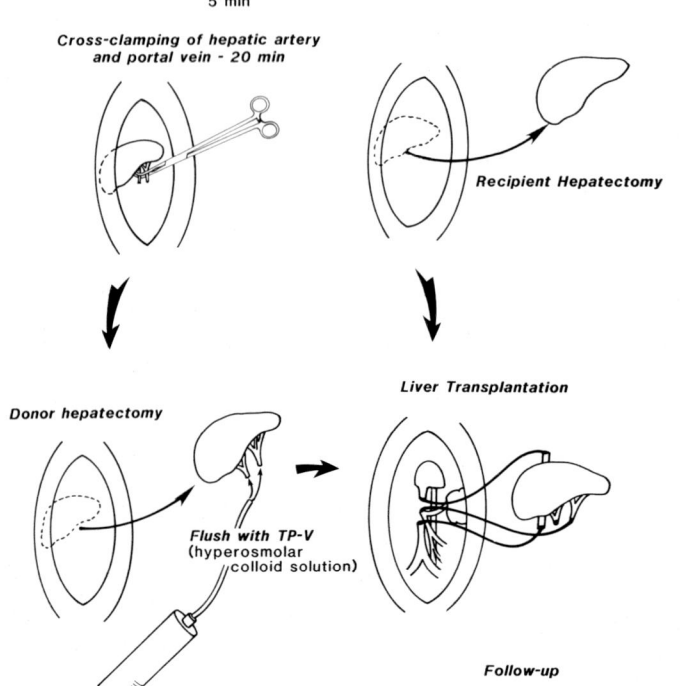

5 min

Cross-clamping of hepatic artery
and portal vein - 20 min

Recipient Hepatectomy

Liver Transplantation

Donor hepatectomy

Flush with TP-V
(hyperosmolar
colloid solution)

Follow-up
2-hour study period –
Laboratory studies, biopsy

Fig. 1. Experimental design of study testing the effects of donor liver pretreatment, with insulin and/or catalase, on the function of liver grafts after ex vivo transplantation.

tery and portal vein were identified. Pretreatment, as previously described was then administered via the portal vein (Gr. II–V). After 5 min, the portal vein and hepatic artery were cross-clamped for 20 min. Following this period, a donor hepatectomy was performed. Recipient animals were anesthetized as described above and hepatectomies were performed. Transplant anastomoses were then accomplished in hepatectomized recipient animals by cannulating ex vivo liver grafts with the portal vein and distal and proximal vena cavas. The hepatic artery was ligated. Follow-up studies were then conducted for 2 h to assess liver function, including: lactic acid dehydrogenase (LDH), serum glutamic oxaloacetic transaminase (SGOT), serum bilirubin and γ-glutamyl transferase (GGT) determinations. pO_2, pCO_2, pH, and liver weight were also measured. After completion of the test period, hepatic biopsy specimens were processed for light microscopy. The data from all groups was statistically compared using the Student's t-test.

Table I. Comparison of laboratory determinations after hepatic transplantation (M±SD)

Group	Treatment	Time h	LDH mg/dl	SGOT IU/l	Total bilirubin mg/dl	GGT U/l
I	control	0	1,881±517	155±21	0.9±0.16	17.2±5.8
		1	3,382±480	342±23	1.5±0.25	25.0±6.6
		2	*	*	*	*
II	insulin	0	687±138[f]	103±25[f]	0.63±0.58[a]	2.1±2.2[f]
	(1,000 U/rat)	1	1,113±857[f]	160±75.3[f]	0.56±0.12[f]	3.1±3.7[f]
		2	1,241±902	193±114	0.62±0.26	3.0±3.7
III	insulin	0	1,211±965[d]	114±53[f]	0.80±1.1	12.6±9.6[b]
	(100 U/rat)	1	1,442±1,269[f]	208±232[c]	0.80±0.77	11.0±10[f]
		2	1,727±1,212	202±150	1.2 ±1.2	33.2±38.5
IV	catalase	0	654±391[f]	160±104	–	18.8±14.5
	(5,000 U/kg)	1	799±242[f]	104±107[f]	–	30.5±23.1
		2	1,083±539	147±83	–	19.5±14.2
V	catalase	0	738±672[f]	102±60[e]	0.46±0.29[f]	2.2±2.0[f]
	(5,000 U/kg)	1	664±581[f]	90±61[f]	0.99±1.0[a]	2.1±1.6[f]
	+ insulin	2	579±665	109±111	0.74±0.64	1.6±0.92
	(1,000 U/rat)					

[a] p < 0.05; [b] p < 0.025; [c] p < 0.01; [d] p < 0.005; [e] p < 0.001; [f] p < 0.0005.

* Rats in control group expired by 2 hour sampling, unable to obtain sample.

Results

Table I details the comparative laboratory results in each of the study groups after transplantation. Initial and 1-hour mean LDH values were significantly decreased (p < 0.005) in all treated groups as compared to controls and decreased or showed only minimal increases by the 2-hour sampling. Initial SGOT levels were significantly (p < 0.001) decreased in all treated groups except group IV, treated with catalase, as compared to control levels. However, at the 1-hour sampling, the SGOT levels of all treated groups were significantly less (p < 0.01) than control values and remained stable to the 2-hour sampling. Initial levels for total bilirubin were significantly less (p < 0.05) in group II, pretreated with high dose insulin and in group V, pretreated with catalase + high dose insulin, than in the non-pretreated control livers. Total bilirubin was also significantly (p < 0.05) decreased in groups II and IV at 1-hour samplings

Table II. Comparative changes in other physical and chemical parameters (M±SD)

Group	Treatment	Weight change* %	Time, h	pH*
I	control	6.52±2.53	0	7.43±0.02
			1	7.35±0.04
			2	7.29±0.05
II	insulin (1,000 U/rat)	3.39±3.96[b]	0	7.42±0.02
			1	7.38±0.03[a]
			2	7.36±0.05[a]
III	insulin (100 U/rat)	6.79±4.34	0	7.44±0.02
			1	7.38±0.02[a]
			2	7.33±0.05[a]
IV	catalase (5,000 U/kg)	1.74±2.45[d]	0	7.44±0.01[a]
			1	7.40±0.01[d]
			2	7.40±0.01[d]
V	catalase (5,000 U/kg) + insulin (1,000 U/rat)	5.23±5.62	0	7.43±0.02
			1	7.38±0.06
			2	7.36±0.09[c]

[a] $p < 0.05$; [b] $p < 0.025$; [c] $p < 0.005$; [d] $p < 0.0005$.
* Student's t-test versus control group.

and remained stable for the next hour after transplantation in these groups. Initial and 1-hour GGT levels were significantly decreased as compared with controls ($p < 0.025$), in groups II, III, and V. However, no significant ($p < 0.05$) difference was observed in GGT between control and group IV values at these times.

Significantly ($p < 0.05$) less mean percent weight change was observed in the livers treated with high dose insulin (Gr. II) or catalase (Gr. IV) (table II).

pH values remained more stable in the treated groups (II–V) than in the untreated control livers (Gr. I) throughout the post-transplant monitoring period (table II). Significant differences ($p < 0.05$) were observed between many of the pH values in the experimental groups versus control livers after transplantation, especially after 2 h.

Histological findings demonstrated hepatic necrosis with severe congestion and obstruction of the portal triad with red blood cells in the control group (fig. 2). Groups II and V showed evidence of protection of hepatic architecture with moderate to minimal congestion and there was no

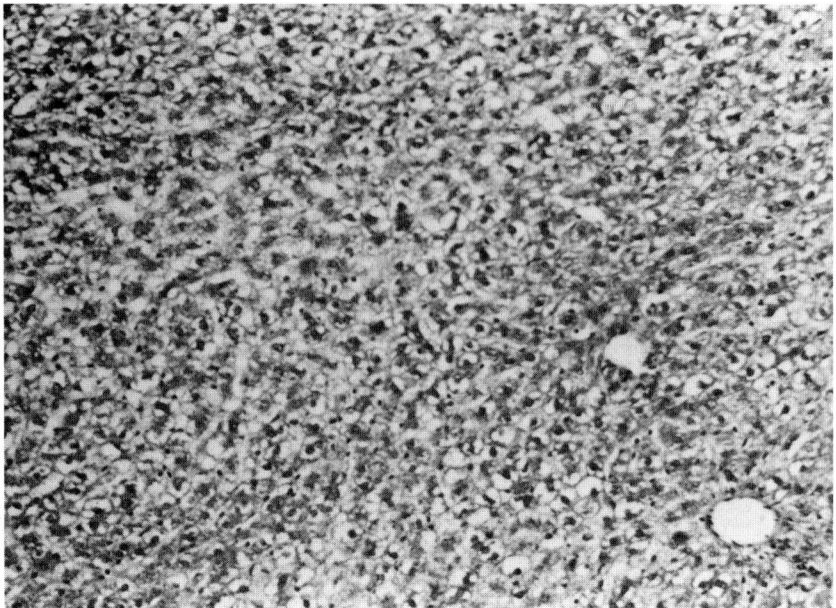

Fig. 2. Total loss of architecture is seen in this liver that did not receive any treatment. Even though the central vein is intact, we cannot identify well-preserved hepatocytes. HE. × 65.

evidence of portal triad obstruction (fig. 3). Groups III and IV had a histological picture only minimally different from controls (fig. 4).

Discussion

The sensitivity of the liver allograft to damage during harvesting and preservation led us to explore new ways of protecting the organ. Studies of the changes occurring during periods of ischemia in various tissues have suggested that the accumulation of oxygen-derived free radicals, such as OH^- and O_2^-, may be responsible for some of the damage observed during organ harvesting and preservation [1]. Recent work in our laboratories on ischemically damaged liver grafts (published in this volume) and experimental investigations by other on ischemic small bowel and heart tissue [4, 5] have indicated that the application of oxygen free radical scavengers, such as catalase and superoxide dismutase, can protect the tissue from ischemic damage and improve functional results after

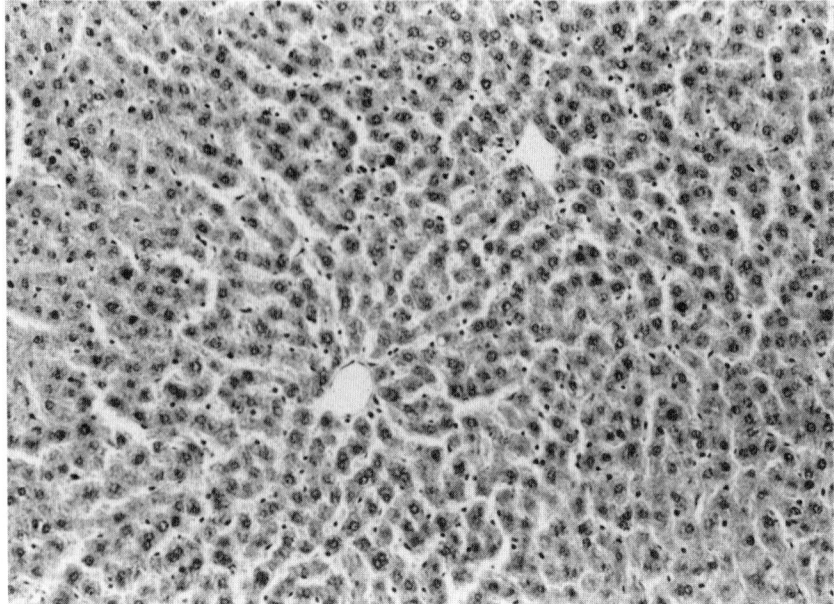

Fig. 3. Moderate to minimal hepatocyte congestion with sinusoidal dilatation is evident in this liver treated with insulin (1,000 U/rat). HE. × 65.

periods of ischemia. Several studies have provided evidence that insulin may act as a hepatotrophic factor [2, 3]. We, therefore decided to test the potential protective effects of catalase and insulin, used alone and in combination, on livers exposed to a period of iatrogenic ischemia. We found that a 20-min cross-clamping of the portal vein and hepatic artery was sufficient to consistently induce hepatic failure in the untreated group. Using this model, catalase and/or insulin were given to the treated groups, prior to induction of ischemia, within a time frame which would be clinically feasible. After donor hepatectomy and subsequent transplantation to hepatectomized recipient animals, the ex vivo transplant set-up provided an opportunity to simultaneously monitor many parameters.

Results of this study indicate that high dose insulin given either alone or with catalase was most beneficial in affecting recovery of the ischemically damaged liver allografts. The livers receiving the lower dosage of insulin or catalase used alone, however, were not similarly protected from damage. It was difficult to determine the exact effects of catalase in group V, when combined with high dose insulin, but probably it

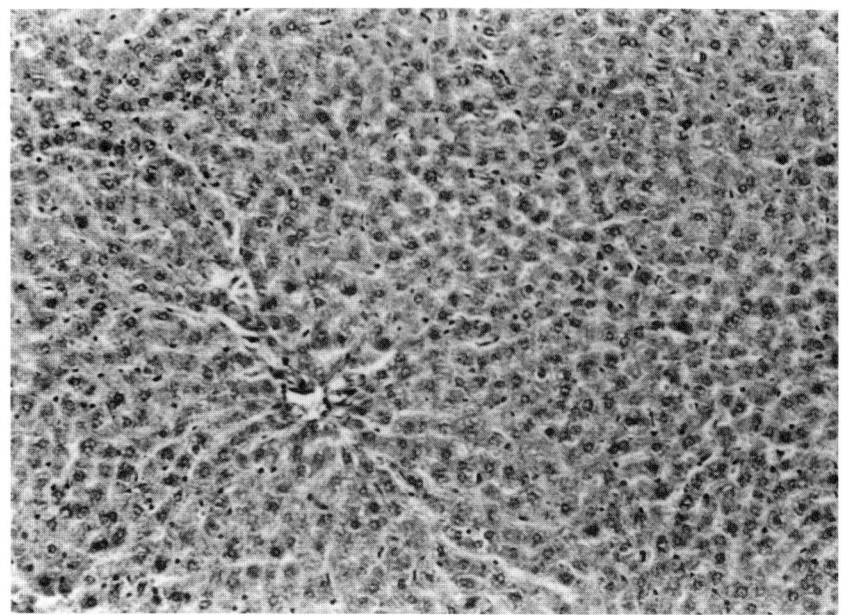

Fig. 4. Near normal hepatic architecture is noted in this liver treated with catalase + insulin (1,000 U/rat). HE. × 65.

is not deleterious to ultimate function. Further studies will elucidate the optimal applications of this work.

References

1   McCord, J. M.: Oxygen-derived free radicals in postischemic tissue injury. New Engl. J. Med. *312*:159–163 (1985).
2   Starzl, T. E.; Terblanche, J.: Hepatotrophic substances; in Popper, Schaffner, pp. 135–151 (Grune & Stratton, New York 1979).
3   Jeejeebhoy, N.; Ho, J.; Mehra, R.; Bruce-Robertson, A.: Hepatotrophic effects of insulin on glucose, glycogen, and adenine nucleotides in hepatocytes isolated from fasted adult rats. Gastroenterology *78:* 556–570 (1980).
4   Granger, D. N.; Rutili, G.; McCord, J. M.: Superoxide radicals in feline intestinal ischemia. Gastroenterology *81:* 22–29 (1981).
5   Shlafer, M.; Kane, P. F.; Kirsh, M. M.: Superoxide dismutase plus catalase enhance the efficacy of hypothermic cardioplegia to protect the globally ischemic, reperfused heart. J. thorac. cardiovasc. Surg. *83:* 830–839 (1982).

L. H. Toledo-Pereyra, MD, Department of Surgery, Section of Surgical Research, Mount Carmel Mercy Hospital, 6071 W. Outer Drive, Detroit, MI 48235 (USA)

Novelli, Ursini (eds.), Oxygen Free Radicals in Shock. Int. Workshop,
Florence 1985, pp. 212–219 (Karger, Basel 1986)

Assessment of Oxygen Free Radical Scavengers on Ischemic Livers

Luis H. Toledo-Pereyra

Department of Surgery, Section of Surgical Research, Mount Carmel Mercy
Hospital, Detroit, Mich., USA

Introduction

The formation of oxygen-derived free radicals has been recently associated with tissue injury during ischemia and reperfusion [1]. Cytotoxic free radicals, including OH^- and O_2^-, accumulate during ischemia because of damage to cellular enzymes which normally eliminate these metabolites. The purpose of this study was to evaluate the efficacy of pretreatment with oxygen free radical scavengers, superoxide dismutase (SOD) and catalase (CAT), in reducing injury to livers subjected to iatrogenic ischemia prior to removal.

Materials and Methods

Adult mongrel dogs, weighing between 14 and 26 kg, were used as liver donors. Animals were heparinized (10,000 U), anesthetized with sodium pentobarbital (20 mg/kg), and oxygen was administered at a rate of 2–3 l/min. Intravenous support consisted of 500 ml 5% dextrose in saline. Donor livers were subjected to 40 min warm ischemia by cross-clamping the portal vein and hepatic artery. Prior to cross-clamping, donors were divided into one untreated control group (Gr. I, n = 6) and four experimental groups receiving intravenous CAT and/or SOD as follows (fig. 1): Gr. II (n = 6) CAT (5,000 U/kg); Gr. III (n = 6) SOD (5,000 U/kg); Gr. IV (n = 6) CAT + SOD (5,000 U/kg, each).

Hepatectomies were performed on all donor dogs after the ischemic period and livers were flushed with cold (4 °C) Ringer's lactate solution through the portal vein 9,400 ml) and hepatic artery (100 ml) until the venous effluent was clear. Prior to placing each organ on the perfusion machine, the portal vein, hepatic artery, and biliary duct were cannulated.

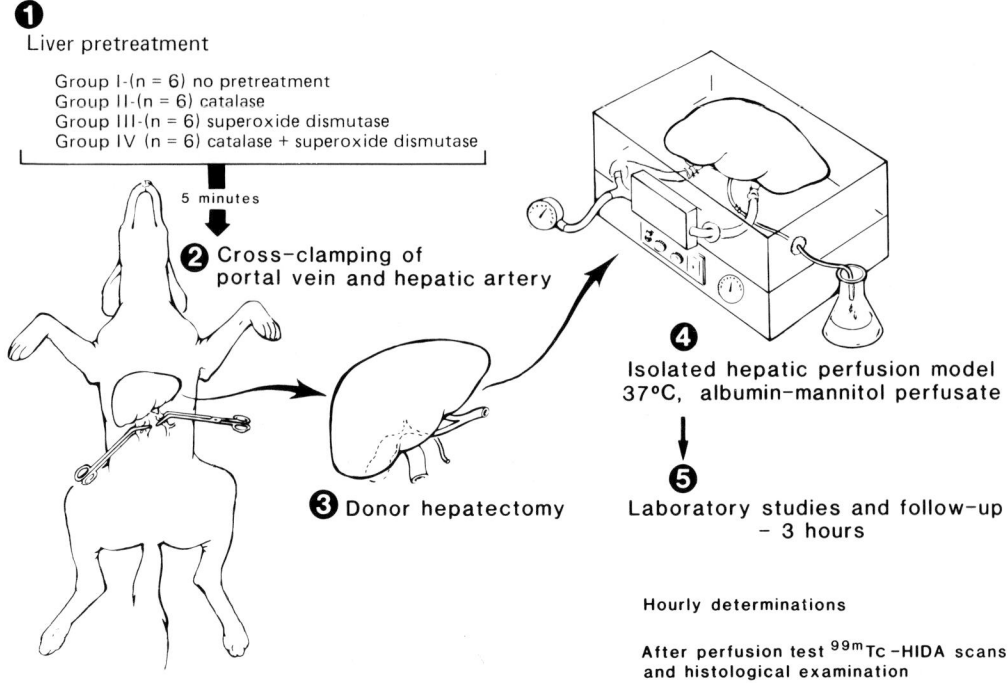

① Liver pretreatment

Group I-(n = 6) no pretreatment
Group II-(n = 6) catalase
Group III-(n = 6) superoxide dismutase
Group IV (n = 6) catalase + superoxide dismutase

5 minutes

② Cross-clamping of
portal vein and hepatic artery

③ Donor hepatectomy

④ Isolated hepatic perfusion model
37°C, albumin-mannitol perfusate

⑤ Laboratory studies and follow-up
- 3 hours

Hourly determinations

After perfusion test ^{99m}Tc-HIDA scans
and histological examination

Fig. 1. Experimental design to test the effects of donor pretreatment with catalase and/or superoxide dismutase to protect ischemic liver grafts.

A normothermic (37 °C) pulsatile perfusion setup was used to evaluate the functional status of each liver for a 3-hour test period. Albumin (Cutter Labs, 500 ml) with mannitol (Abbott, 25 mg/l) was used as the perfusate on the modified MOX-100 machine (Waters Instruments, Rochester, Minn.). Hepatic artery pressure was maintained up to 100 mm Hg and portal pressure was kept at 8–10 cm H_2O.

Functional status was monitored at hourly intervals by collecting bile and by obtaining perfusate samples for serum glutamic oxaloacetic transaminase (SGOT), lactic acid dehydrogenase (LDH), alkaline phosphatase, γ-glutamyl transferase (GGT), total bilirubin, pO_2, pCO_2, pH and lactic acid. Portal venous flow and hepatic artery flow were also monitored at hourly intervals. Student t-tests was used for comparisons between mean values of control and experimental groups.

After the perfusion test period, ^{99m}Tc-HIDA scans were performed to assess the vascular and functional integrity of the livers. Stannous technetium-disofenin (Hepatolite, New England Nuclear, 5 mCi) was injected into the additive port of the bubble trap prior to liver perfusion recordings at a rate of 1/s. After completion of the HIDA scans, biopsy specimens were submitted for light microscopic histological evaluation.

Table I. Comparative changes for some laboratory values in the control and experimental groups

Group	LDH	SGOT	AP	GGT	Lactic acid
I – Control	+++	+++	+++	+	++
II – CAT	++	+	++	+	++
III – SOD	+	+	+++	+	++
IV – CAT + SOD	++	+	++	+	++

CAT = Catalase; SOD = superoxide dismutase; +++ = severe increase; ++ = moderate increase; + = minimal increase; LDH = lactic acid dehydrogenase, +++ ($>1,200$ IU/l), ++ (200–1,200 IU/l), + (20–200 IU/l); SGOT = serum glutamic oxaloacetic transaminase, +++ (>300 IU/l), ++ (150–300 IU/l), + (20–150 IU/l); AP = alkaline phosphatase, +++ (40–100 mM/ml), ++ (10–40 mM/ml), + (<10 mM/ml); GGT = γ-glutamyl transferase, + (1–6 U/l); lactic acid, +++ (>15 mM/l), ++ (9–15 mM/l), + (3–8 mM/l).

Results

During the isolated perfusion test period, the most notable changes in LDH, SGOT in the treated groups as compared with controls were observed in Gr. II, treated with CAT and Gr. IV, treated with CAT + SOD (table I). Only minimal differences were seen in the perfusate alkaline phosphatase, lactic acid, and GGT levels measured in each group. Similarly, initial mean pH values for the four groups were not significantly different ($p > 0.05$; range of means, 7.43–7.45). A significant ($p < 0.0005$) decrease in pH, however, was noted in the control group by the end of the 3-hour perfusion test period (M±SD = 7.04±0.07) as compared to pH levels in the other groups (Gr. II, 7.31±0.02; Gr. III, 7.31±0.02; Gr. IV, 7.38±0.02).

Fig. 2. Light histology of control and experimental groups of ischemic livers. *a* There is moderate sinusoidal dilatation with minimal to no congestion. This liver treated with CAT + SOD has near normal histology. HE. × 70. *b* Impending necrosis of the liver parenchyma is shown at 3 to 6 o'clock. Focal dilatation of sinusoidal spaces is also present in this liver treated with SOD. HE. × 70. *c* There is moderate congestion and sinusoidal dilatation with no obliteration of the central vein in this liver treated with CAT. HE. × 70. *d* There is disorganization and moderate loss of hepatic architecture. The hepatocytes are swollen with retained red blood cells in the portal triad and appearance of fibrosis-like changes in this control ischemic liver. HE. × 70.

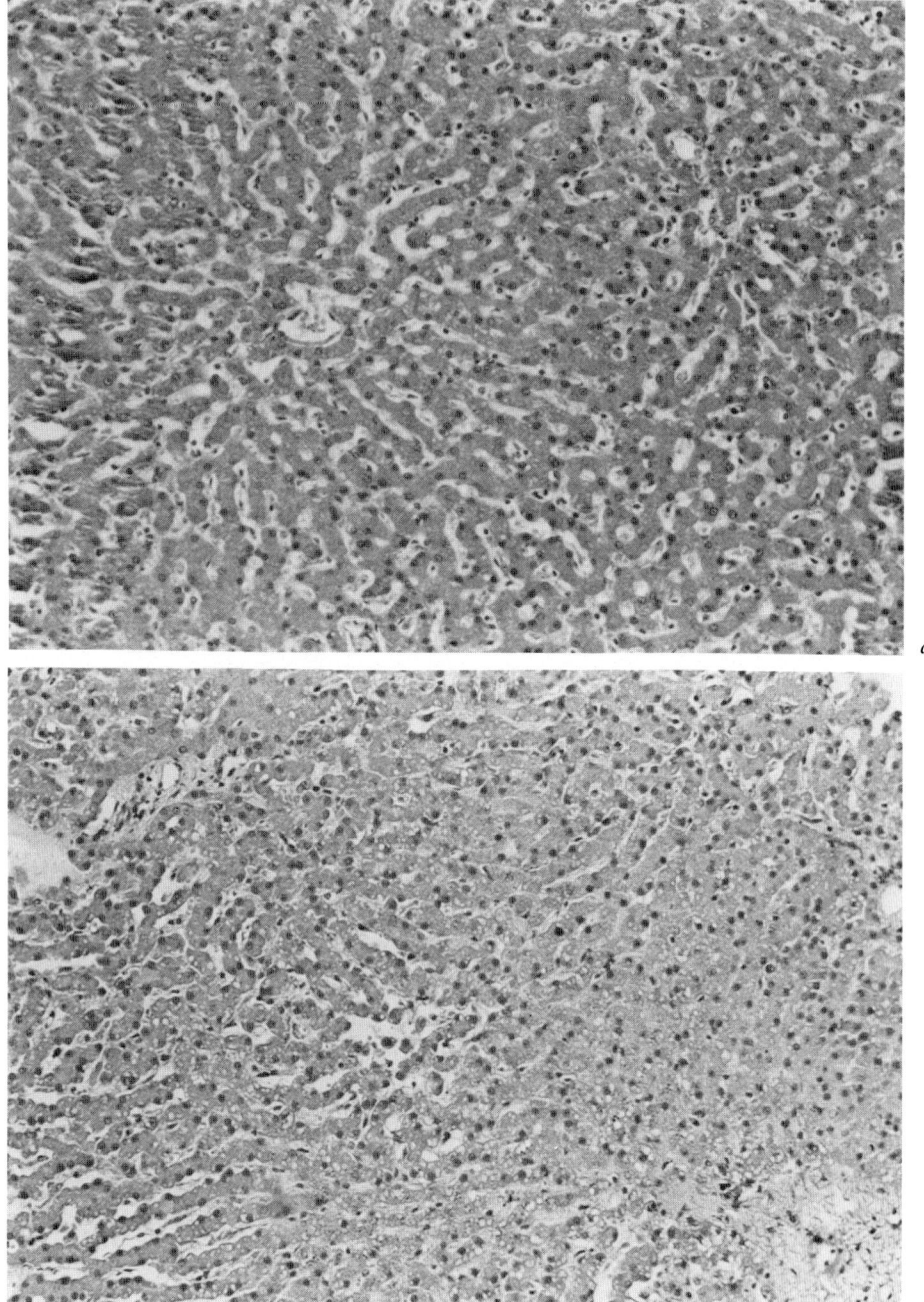

a

b

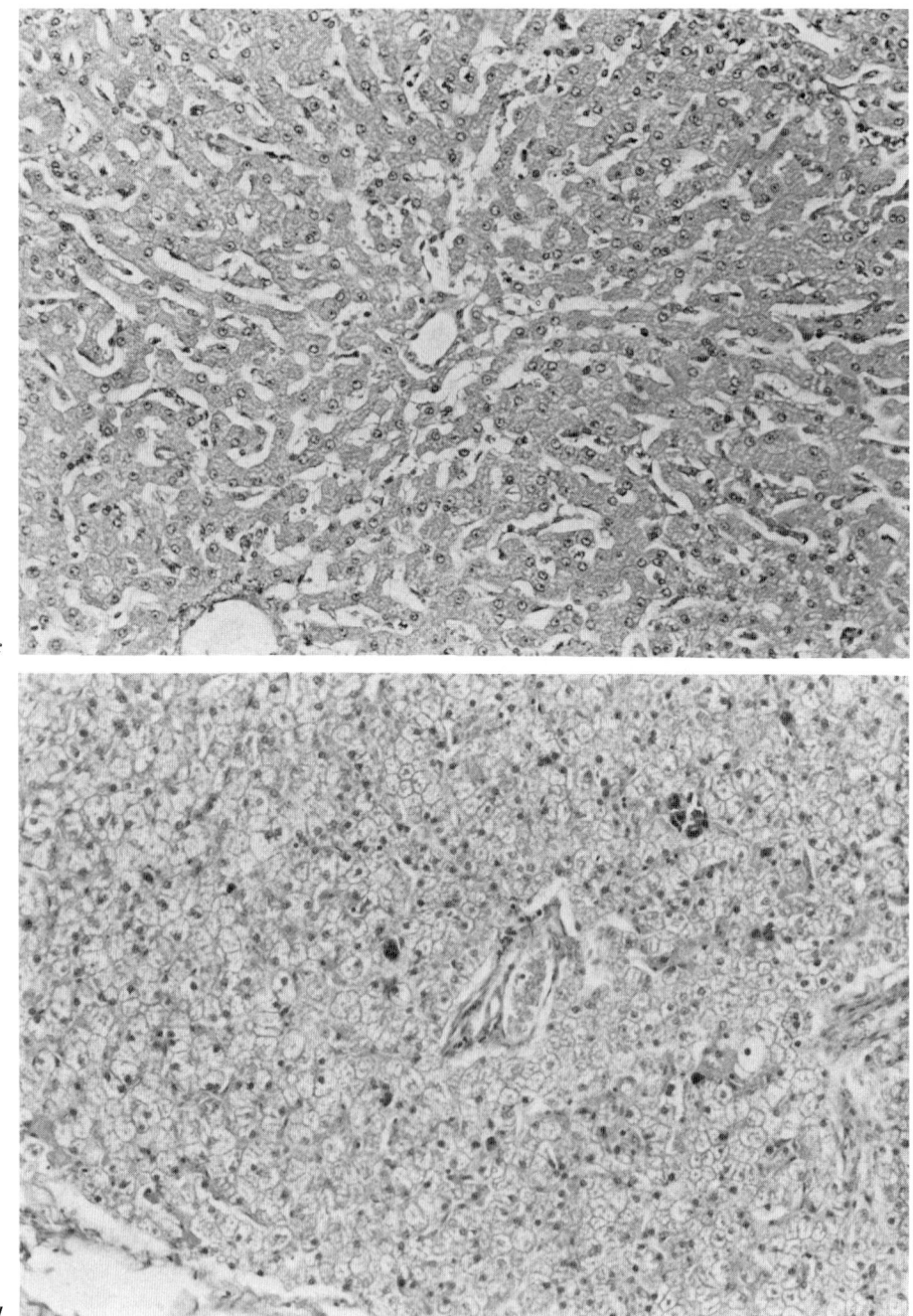

c

d

Fig. 2c, d (for legend see p. 214)

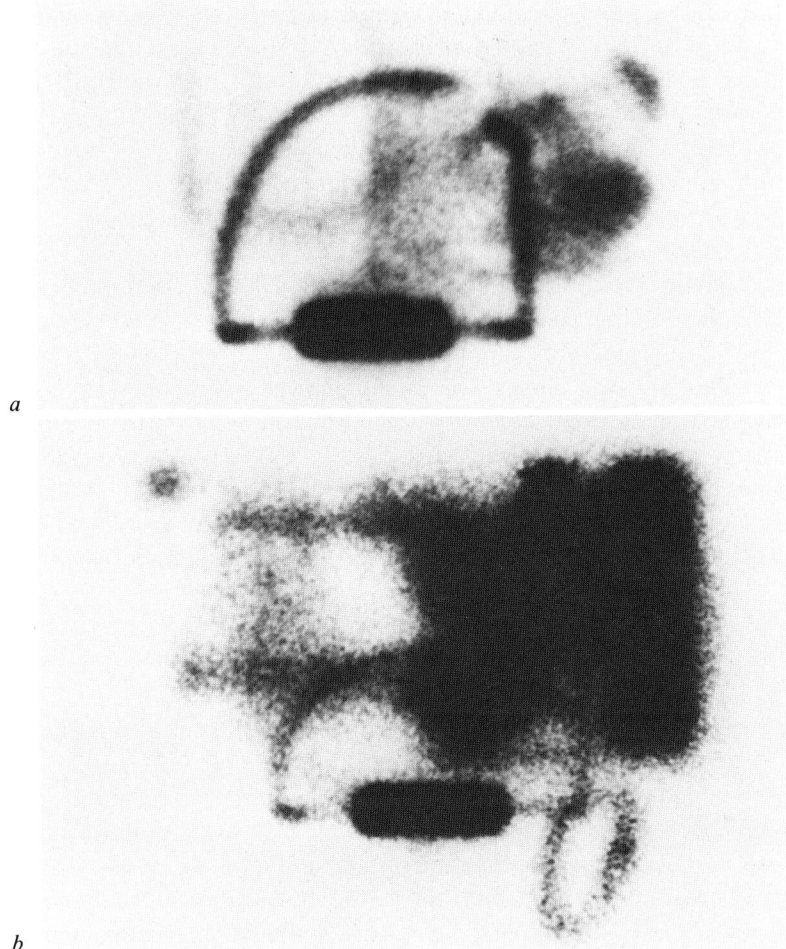

a

b

Fig. 3. ⁹⁹ᵐTc-HIDA scans of selected livers: *a* Poor uptake and perfusion were observed in this frame from a HIDA scan taken of a control liver (Gr. I) at 36 s after injection of the dye. *b* HIDA image of CAT + SOD-treated liver (Gr. IV) taken at same frame time as *a*. Improved perfusion and vascular integrity is evident, as well as bile production (upper left of figure).

Although there were no significant ($p > 0.05$) differences in the portal venous (PV) and hepatic arterial (HA) flows between the groups at the initiation of the isolated perfusion test (range of mean flows, PV = 233–259 ml/min, HA = 136–145 ml/min), the improved vascular function was evidenced by significant increases in both PV and HA flows in all treated groups as compared with controls by the 3-hour sampling (3 h control PV = 126.5±49.8 ml/min, HA = 89.7±11 ml/min; Gr. II – PV = 255±34.3 ml/min, HA = 141.3±42 ml/min; Gr. III – PV = 187.2±17.9 ml/min, HA = 107.7±6.9 ml/min; Gr. IV – PV = 258±15.7 ml/min, HA = 133±8.2 ml/min).

Initial mean bile flows were significantly (p < 0.001) increased in Gr. II (10±3.4 ml/h) and Gr. IV (8.3±1.9 ml/h) as compared with control (3±1.7 ml/h) and Gr. III (3.7±1.0 ml/h) levels. However, all experimental groups demonstrated significantly (p < 0.0005) better bile flows at the 3-hour sampling (Gr. II, 12.3±2.2 ml/h; Gr. III, 3.5±1.0 ml/h, and Gr. IV, 6.8±2.2 ml/h) as compared with control levels (0.66±1.2 ml/h).

Studies of liver biopsy specimens showed improved histological characteristics in livers receiving CAT and/or SOD pretreatments as compared to control specimens (fig. 2a–d). ^{99m}Tc-HIDA scan results indicated improved perfusion, uptake and biliary excretion in Gr. II, treated with CAT and Gr. IV, treated with CAT + SOD, as compared with Gr. I controls and Gr. III livers treated with SOD alone (fig. 3a, b).

Discussion

Studies of organs subjected to periods of ischemia have indicated that the abundant production of oxygen-free radicals, OH^- and O_2^- may play an important role in tissue damage [1]. During ischemia and reperfusion, this increase in oxygen-derived free radicals has been linked to increases in the enzyme, xanthine oxidase [2] and a failure of catalase and superoxide dismutase to convert the cytotoxic free radicals to less harmful metabolites [2]. Studies of the effects of ischemia on intestinal mucosa and myocardium have indicated the value of supplementing the tissue with exogenous oxygen free radical scavengers, catalase and superoxide dismutase [3, 4]. Xanthine oxidase inhibitors, such as allopurinol, and other graft and donor pretreatment modalities have also been utilized to improve post-transplant function of various grafts [5–9].

In our present study, oxygen free radical scavengers have been used to improve the function of livers previously exposed to prolonged ischemia. Using an isolated perfusion model, which allowed for simultaneous testing of multiple variables, the results indicate that pretreatment of donor livers with catalase and/or superoxide dismutase, prior to ischemia yielded improved functional status and better vascular integrity after donor hepatectomy. This was demonstrated by laboratory determinations, physical parameters, histological findings and HIDA scans. The results of this work may provide new alternatives for protection of livers from the effects of ischemia and reperfusion during preservation prior to transplantation.

References

1   McCord, J. M.: Oxygen-derived free radicals in postischemic tissue injury. New Engl. J. Med. *312:* 159–163 (1985).
2   McCord, J. M; Fridovich, I: The reduction of cytochrome *c* by milk xanthine oxidase. J. biol. Chem. *243:* 575–560 (1968).
3   Shlafer, M; Kane, P. F.; Kirsh, M. M.: Superoxide dismutase plus catalase enhance the efficacy of hypothermic cardioplegia to protect the globally ischemic, reperfused heart. Cardiovasc. Surg. *83:* 830–839 (1981).
4   Owens, M. L.; Lazarus, H. M.; Wolcott, M. W.; et al.: Allopurinol and hypoxanthine pretreatment of canine kidney donors. Transplantation *17:* 424–427 (1974).
5   Toledo-Pereyra, L. H.; Simmons, R. L.; Najarian, J. S.: Effect of allopurinol on the preservation of ischemic kidneys perfused with plasma or plasma substitutes. Ann. Surg. *180:* 780–782 (1974).
6   Toledo-Pereyra, L. H.; Najarian, J. S.: Effective treatment of severely damaged kidneys prior to transplantation. I. Preliminary results. Transplantation *16 D:* 63 (1973).
7   Toledo-Pereyra, L. H.; Simmons, R. L.; Najarian, J. S.: Comparative effects of chlorpromazine, methylprednisolone, and allopurinol during small bowel preservation. Am. J. Surg. *126:* 631 (1973).
8   Toledo-Pereyra, L. H.; Simmons, R. L.; Najarian, J. S.: Protection of the ischemic liver by donor pretreatment before transplantation. Am. J. Surg. *129:* 129 (1975).
9   Toledo-Pereyra, L. H.: Restoration of function after ischemia injury on kidneys treated with allopurinol; PhD thesis, Minneapolis (1976).

L. H. Toledo-Pereyra, MD, Department of Surgery, Section of Surgical Research, Mount Carmel Mercy Hospital, 6071 W. Outer Drive, Detroit, MI 48235 (USA)

Novelli, Ursini (eds.), Oxygen Free Radicals in Shock. Int. Workshop,
Florence 1985, pp. 220–223 (Karger, 1986)

Augmenting Effects of Cerebral Ischaemia on the Systemic Response to Haemorrhagic Hypotension in the Rat

Bjarne Grögaard, Bengt Gerdin, Karl-E. Arfors

Department of Experimental Medicine, Pharmacia AB, and Department of
Surgery, University Hospital, Uppsala, Sweden

Introduction

After multitrauma cerebral ischaemia, arterial hypotension and
haemorrhagic shock occur simultaneously in certain patients. Cerebral ischaemia and systemic hypotension are two of the most powerful activa-
tors of the sympathetic vasoconstrictor system. It is therefore reasonable
to suppose that the cerebral ischaemic response could potentiate the systemic effect of haemorrhagic hypotension. In the present study this question was investigated in a rat model originally developed for studies on
cerebral ischaemia.

Materials and Methods

Fasted male Wistar rats, 300–400 g, were used. They were intubated and maintained on N_2O/O_2 anaesthesia under controlled ventilation. Catheters were inserted
into the tail artery and vein, into the right atrium, into the left femoral artery and via
the right brachial artery into the left ventricle. The carotid arteries were dissected free
and prepared for clamping.

Blood Flow Measurements

Cardiac output and organ blood flows were determined by a radioactive micro-
sphere technique [*Lundberg and Smedegård,* 1981]. 0.4×10^6 spheres, $\varnothing$ 15 µm, labelled
with ^{95}Nb, ^{85}Sr or ^{65}Zn were injected into the left ventricle.

The first group of animals (n = 6; bleeding + clamping; BC) were bled through
the central venous catheter until a mean arterial blood pressure (MAP) of 80 mm Hg
was attained, the two common carotid arteries were clamped and bleeding was rapidly

continued to MAP of 50 mm Hg. After 15 min of ischaemia, the shed blood was rein-fused and at MAP of 80 mm Hg the clamps were released. The animals were then al-lowed a 60-min recirculation period. The second group of animals (n = 7; bleeding; B) were identically prepared, and bled to an MAP of 50 mm Hg without clamping of the carotid arteries. In the third group (n = 8; sham-operated) the animals were operated on in the same manner as above but ischaemia and haemorrhage were not instituted.

Blood flow determinations were performed at the end of the steady-state period, after 12 min of ischaemia and after 60 min of recirculation. The acid-base status, blood glucose values and haematocrit were measured before ischaemia, at the end of the is-chaemic period and at 5, 30 and 60 min of recirculation.

Results

More profound bleeding was required to obtain MAP of 50 mm Hg in group BC than in group B (2.6 ± 0.1 vs 2.2 ± 0.4 mm/100 g body weight, $p < 0.05$), which is indicative of a potentiated vasoconstrictor response in the former group. During hypotension there was more profound meta-bolic acidosis in group BC than in group B (fig. 1; pH 7.16 ± 0.06 and BE -17.2 ± 4.1 vs pH 7.37 ± 0.01 and BE -11.3 ± 4.3). The metabolic acidosis was still uncompensated after 60 min of recirculation in group BC (pH 7.18 ± 0.08 and BE -14.5 ± 3.6), but was better corrected in group B (pH 7.35 ± 0.02 and BE -5.2 ± 2.3). The hypotension was followed by an in-crease in blood glucose in group B but not in group BC (fig. 1). After 60 min of recirculation the blood glucose concentration had decreased in group BC compared with the steady-state value, but remained unaltered in group B.

Cardiac output decreased similarly in the two experimental groups during hypotension to about $\frac{1}{5}$ of the steady-state value and correspond-ingly there was an increase in total peripheral resistance to about twice the steady-state value. Animals subjected to both bleeding and clamping (group BC) exhibited a decrease in renal blood flow to 6% of the steady-state value and to 18% in animals subjected to bleeding alone (group B; fig. 2).

There was a considerable increase in renal resistance in both groups, which was significantly more profound in group BC. In group BC the blood flow in the spleen and skin virtually ceased while these flows were better preserved in animals subjected to bleeding only (group B; fig. 2). The blood flow in the liver was slightly more deranged in group BC than in group B during hypotension, and was significantly lower after 60 min of recirculation in this group (fig. 2).

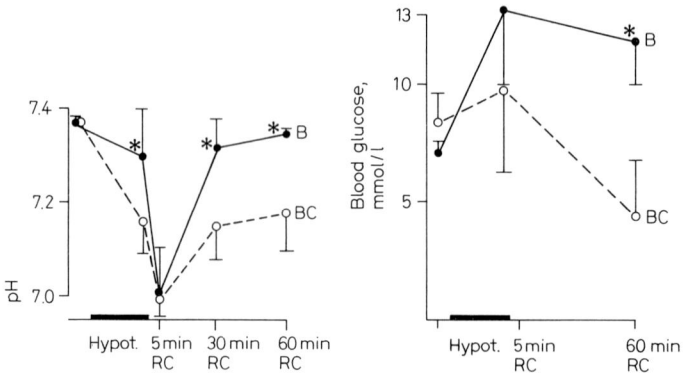

Fig. 1. pH and blood glucose concentrations in groups B and BC before, during and after hypotension. * p < 0.05. Mean±SD.

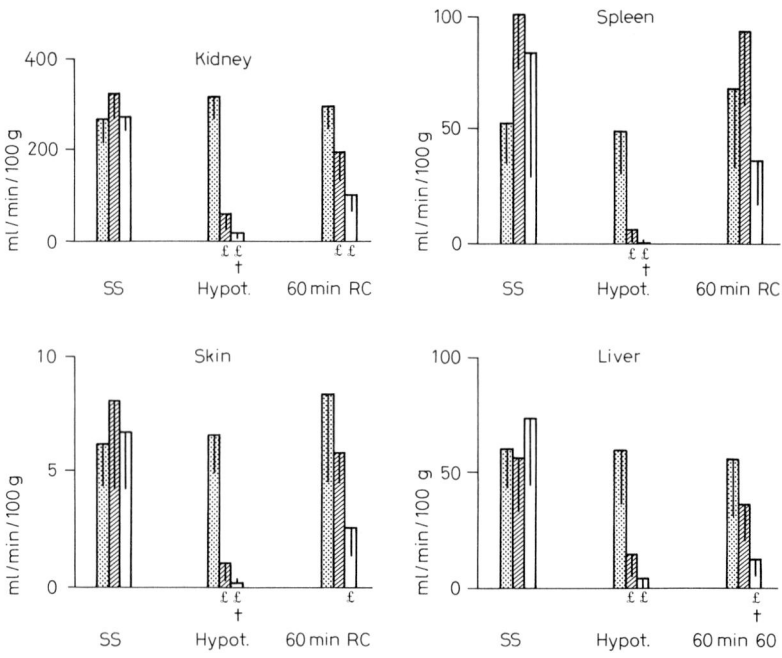

Fig. 2. Blood flows in the kidney, spleen, skin and liver before, during and 60 min after hypotension in sham-operated animals, and in groups B and BC. Mean±SD. £=p < 0.05 vs sham; †=p < 0.05 vs B.

Discussion

The present study indicates that carotid artery occlusion potentiates the vascular response to haemorrhage and also alters the hypotension-induced mobilization of tissue glycogen, leading to a normal or low blood glucose concentration during and after hypotension.

The potentiated vasoconstrictor response can be explained simply by an increased release of vasoconstrictor amines. The extinction of the hypotension-induced hyperglycaemia by the cerebral ischaemia, however, is more difficult to explain. As the ability of animals to withstand and to survive haemorrhagic shock is related to their ability to maintain a hyperglycaemic state [e.g. *Strawitz* et al., 1961], it is possible that this effect is involved in the unexpectedly high mortality observed after 12–13 min of carotid artery clamping + arterial hypotension in a previous study [*Grögaard* et al., in press]. The mechanism by which cerebral ischaemia interferes with the hypotension-induced hyperglycaemia is presently the subject of further investigation.

References

Grögaard, B.; Gerdin, B.; Arfors, K.-E.: Forebrain ischaemia in the rat. Relation between duration of ischaemia, use of adjunctive ganglionic blockade and long-term recovery. Stroke (in press).

Lundberg, C.; Smedegård, G.: Regional differences in skin blood flow as measured by radioactive microspheres. Acta physiol. scand. *111:* 491–496 (1981).

Strawitz, J. G.; Hift, H.; Ehrhardt, A.; Cline, D. W.: Irreversible hemorrhagic shock in rats: changes in blood glucose and liver glycogen. Am. J. Physiol. *200:* 261–263 (1961).

B. Grögaard, MD, Department of Experimental Medicine, Pharmacia AB, Uppsala (Sweden)

Novelli, Ursini (eds.), Oxygen Free Radicals in Shock. Int. Workshop,
Florence 1985, pp. 224–230 (Karger, Basel 1986)

α-Tocopherol, Mannitol and Methylprednisolone Prevention of FeCl₂ Initiated Free Radical Induced Lipid Peroxidation in Spinal Cord[1]

Douglas K. Anderson[a], Eugene D. Means[b]

[a] Research and [b] Neurology Services, Cincinnati Veterans Administration Medical
Center, [a,b] Departments of Neurology and [a] Physiology, University of Cincinnati
College of Medicine, Cincinnati, Ohio, USA

Introduction

The purpose of this study was 3-fold. The first was to determine in vivo the effects of iron, a known free radical generator, on cellular membranes by microinfusing $FeCl_2$ into the spinal cord of cats. Activity of the plasma membrane-bound enzyme, sodium, potassium-activated adenosine triphosphatase (Na^+, K^+-ATPase) was assayed as an index of the occurrence of lipid peroxidation [1, 2] and, hence, as an indicator of plasma membrane integrity. Measurement of thiobarbituric acid-reactive substances (TBARS) and effects of treatment with the antioxidants α-tocopherol and selenium on Na^+, K^+-ATPase activity and TBARS production were used to confirm the occurrence of lipid peroxidation in $FeCl_2$-injected spinal cords.

The second objective of this study was to implicate the hydroxyl radical (.OH) as the reactive oxygen species responsible for initiating the membrane lipoperoxidation. To accomplish this, the ability of mannitol, a purported specific .OH trap [3–8], to reverse the effects of $FeCl_2$ infusion on spinal cord Na^+, K^+-ATPase activity and TBARS production was determined.

The final objective of this study was to use this model of in vivo iron-induced lipid peroxidation to assess the antioxidant potential of methyl-

[1] This work was supported by the Veterans Administration.

prednisolone sodium succinate (MPSS). This synthetic glucocorticoid has been shown to reduce lipid peroxidation in liposomes, thereby demonstrating its antioxidant capability in vitro [9].

Materials and Methods

All of the operative procedures and biochemical analyses utilized for this study have been described in detail in previous publications [1, 10]. Briefly, 29 gauge stainless steel needles were stereotoxically inserted into both the left and right anterior horns of spinal cords at the level of the fourth lumbar vertebra (L_4) in pentobarbital anesthetized cats. 5 µl of either 100 mM $FeCl_2$ (experimental) or 0.9% (w/v) NaCl (control) was infused over a 5-min period into each anterior horn for a total of 10 µl per spinal cord. In one group of cats, the $FeCl_2$ or NaCl was dissolved in a 1% (55 mM) mannitol solution and 5 µl of the resultant solution was infused into each anterior horn. At different times following infusion, the L_4 spinal cord segment was frozen in situ with liquid nitrogen and excised. 'Normal' values for Na^+, K^+-ATPase activity and TBARS levels were obtained on spinal cords frozen in situ through an intact vertebral column (i.e. the spinal cord had not previously been exposed by laminectomy).

Activity of Na^+, K^+-ATPase or concentration of TBARS was determined on 18–20 mg of the gray matter core from each injected L_4 spinal cord segment. The rate of liberation of Pi by the enzymatic hydrolysis of adenosine triphosphate (ATP) was used to determine ATPase activity [11]. Activity of Na^+, K^+-ATPase was the difference in the rate of ATP hydrolysis with and without 9 mM ouabain in the incubation media. Na^+, K^+-ATPase activity was measured at either 1, 2, 8, 24 or 72 h postinfusion and is expressed as nmol Pi/min/mg protein.

Concentration of TBARS (formerly called 'malondialdehyde') was determined by reacting these aldehydic by-products of polyunsaturated fatty acids (PUFA) with thiobarbituric acid [12, 13]. Levels of TBARS were determined at 2 h after either $FeCl_2$ or NaCl infusion and are expressed as pmol/mg protein.

Antioxidant-treated animals were given 1,000 IU of α-tocopherol and 25 µg of selenium p.o. daily for 5 days prior to injection. This treatment regimen was continued until the time of sacrifice.

Glucocorticoid-treated cats were given intravenously (i.v.) 30 mg/kg MPSS 30 min after intraspinal infusion of either NaCl or $FeCl_2$. An additional 15 mg/kg MPSS was given i.v. at both 2.5 and 5.5 h postinfusion. 45 mg/kg MPSS was then infused i.v. in 10 ml of 0.9% NaCl over the next 18 h. The total MPSS dose for the 24-hour period following intraspinal NaCl or $FeCl_2$ infusion was 105 mg/kg.

Results

Following intraspinal infusion of NaCl, Na^+, K^+-ATPase activity had declined to about 65% of normal by 8 h and remained at this level for at least 3 days (table I). Pretreatment of NaCl-infused cats with α-toc-

Table I. Effects of $FeCl_2$, AT and Se, mannitol, and MPSS on spinal cord Na^+,K^+-ATPase activity (nmol Pi/min/mg protein)

Time, h	'Normal' (no laminectomy or infusion)	NaCl (control)	NaCl+ AT and Se	NaCl+ mannitol	NaCl+ MPSS	FeCl₂	FeCl₂+ AT and Se	FeCl₂+ mannitol	FeCl₂+ MPSS
0	125±5								
1		141±11			217±27	131±10			126±13
2		140±4	82±2	143±14	192±16	39±5	45±3	95±11	130±6
8		80±6	87±7			48±5	59±1		
24		76±7	76±9	139±10	127±3	29±7	73±4	100±5	107±2
72		74±6	114±14			47±5	86±8		

MPSS = Methylprednisolone sodium succinate; AT = α-tocopherol; Se = selenium.

opherol and selenium had little effect on Na^+,K^+-ATPase activity for the first 24 h. However, by 3 days, spinal cord Na^+,K^+-ATPase activity had normalized in NaCl infused, α-tocopherol and selenium-pretreated cats (table I). Infusion of NaCl in 1% mannitol into the spinal cord did not change Na^+,K^+-ATPase activity from normal values for the first 24 h following infusion (table I). Treatment with MPSS significantly elevated activity of Na^+,K^+-ATPase above prelaminectomy levels for at least the first 2 h following NaCl infusion. By 24 h, spinal cord Na^+,K^+-ATPase activity had returned to normal levels in these NaCl-infused, MPSS-treated cats (table I).

Infusion of $FeCl_2$ had no effect on spinal cord Na^+,K^+-ATPase for the first hour (table I). However, by the second hour, Na^+,K^+-ATPase activity had declined to 31% of normal and remained at approximately this level for at least 72 h. Infusion of $FeCl_2$ in a 1% mannitol solution statistically negated the effect of $FeCl_2$ on spinal cord Na^+,K^+-ATPase activity (table I). Similarily, treatment with MPSS prevented the effects of intraspinal infusion of $FeCl_2$ on Na^+,K^+-ATPase activity for at least 24 h postinfusion.

Intraspinal infusion of NaCl had no effect on levels of TBARS at 2 h whereas infusion of $FeCl_2$ resulted in levels of TBARS to be elevated almost 2-fold at the same time period (table II). Both infusion of $FeCl_2$ in 1% mannitol and treatment of $FeCl_2$ infused cats with MPSS statistically prevented the $FeCl_2$-induced increase in spinal cord TBARS.

Table II. Level of spinal cord TBARS 2 h after intraspinal NaCl or FeCl$_2$ infusion

Conditions	TBARS concentration pmol/mg protein
'Normal' (no laminectomy or infusion)	6.0±0.05
NaCl infusion (control)	5.7±0.5
NaCl + mannitol	6.8±0.9
NaCl + MPSS	2.8±0.2
FeCl$_2$ infusion	10.7±1.0
FeCl$_2$ + AT and Se	11.5±1.4
FeCl$_2$ + mannitol	7.6±0.5
FeCl$_2$ + MPSS	6.7±0.03

MPSS = Methylprednisolone sodium succinate; AT = α-tocopherol; Se = selenium.

Discussion

In this study, the integrity of plasma membranes was assessed by measuring the activity of Na$^+$, K$^+$-ATPase. Na$^+$, K$^+$-ATPase is a membrane-bound, phospholipid-dependent enzyme whose activity depends on an intact membrane structure [14–17]. The membrane phospholipids appear to be involved in determining the active confirmational structure of this enzyme [14, 15]. Thus, alterations in the structure or form of membrane phospholipids can cause changes in the configuration of Na$^+$, K$^+$-ATPase resulting in a decline in enzymatic activity [14, 15]. Free radical attack of the PUFA of membrane phospholipids will significantly alter the structure of these phospholipids [18]. Hence, a decline in Na$^+$, K$^+$-ATPase activity is one indication of lipid peroxidation in plasma membranes.

Infusion of FeCl$_2$ into the spinal cord of cats caused an increase in the levels of TBARS and a partial inactivation of Na$^+$, K$^+$-ATPase. Histologically, infusion of FeCl$_2$ caused marked necrosis of the spinal cord [1]. Ultrastructurally, plasma membranes and mitochondria contained electron-dense particles of iron and both cellular structures were frequently fragmented and disrupted [10]. Pretreatment of FeCl$_2$-infused cats with α-tocopherol and selenium completely prevented the tissue necrosis [10] and restored Na$^+$, K$^+$-ATPase activity to control (NaCl infused) levels. In aggregate, these findings suggest that iron-initiated free radical reactions in membrane lipids significantly disrupt these membranes causing loss of enzymatic activity and tissue necrosis.

In aerobic aqueous solutions, Fe^{2+} is oxidized to Fe^{3+} by H_2O_2 producing the .OH by the iron catalyzed Haber-Weiss reaction [19–21]. The .OH is a powerful oxidant capable of initiating lipid peroxidation [3, 22, 23]. Mannitol (which has been used extensively as an .OH trap [3–7]) and whose rate constant for its reaction with .OH justifies its use as an .OH scavanger [8] prevented the iron-induced increase in levels of TBARS and decrease in Na^+, K^+-ATPase activity implicating the .OH as the active oxidizing agent of membrane PUFA in this model. However, the nonspecificity of many of the compounds used as specific .OH traps has been noted [6, 24]. Alternatively, a prooxidant with .OH-like reactivity generated by the infused iron may be the oxidizing species responsible for the peroxidation of membrane PUFA.

Reversal of the $FeCl_2$-induced partial inactivation of Na^+, K^+-ATPase and increase in level of TBARS by MPSS is demonstration of the considerable antioxidant (or antioxidant-like) capabilities of this synthetic glucocorticoid at the dosages used in this study. How the MPSS molecule functions as an antioxidant is not known. *Braughler* [25] and *Hall* [26] have recently provided evidence suggesting that the locus for antioxidant activity on the MPSS molecule is the 1,2 double bond [26]. Perhaps the 6-α-methyl moiety on the MPSS molecule in some way increases the antioxidant capacity of this steroid, for example, by increasing its lipid solubility and hence its ability to intercalate into membranes.

In summary, infusion of $FeCl_2$ into the spinal cord of cats appears to damage cell membranes by the process of free radical-induced lipid peroxidation. Reversal of this effect by mannitol implicates the .OH (or an oxidizing agent with .OH-like reactivity) as the initiating radical species in this model of lipid peroxidation. Prevention of this iron catalyzed lipid peroxidation by MPSS suggests a substantial antioxidant (or antioxidant-like) capacity for this glucocorticoid.

References

1   Anderson, D. K.; Means, E. D.: Iron-induced lipid peroxidation in spinal cord: protection with mannitol and methylprednisolone. J. Free Rad. Biol. Med. *1:* 59–64 (1985).
2   Kovachich, G. B.; Mishra, O. P.: Partial inactivation of Na^+, K^+-ATPase in corti-

cal brain slices incubated in normal Krebs-Ringer phosphate medium at 1 and 10 atm oxygen pressures. J. Neurochem. *36:* 333–335 (1981).

3   Fong, K. L.; McCay, P. B.; Poyer, J. L.; Keele, B. B.; Misra, H.: Evidence that peroxidation of lysosomal membranes is initiated by hydroxyl free radicals produced during flavin enzyme activity. J. biol. Chem. *248:* 7792–7797 (1973).

4   Tien, M.; Svingen, A.; Aust, S. D.: Superoxide-dependent lipid peroxidation. Fed. Proc. *40:* 179–182 (1981).

5   Tien, M.; Svingen, A.; Aust, S. D.: An investigation into the role of hydroxyl radical in xanthine oxidase-dependent lipid peroxidation. Archs Biochem. Biophys. *216:* 142–151 (1982).

6   Dillard, C. J.; Kunert, K. J.; Tappel, A. L.: Effect of vitamin E, ascorbic acid and mannitol on alloxan-induced lipid peroxidation in rats. Archs Biochem. Biophys. *216:* 204–212 (1982).

7   Hillered, L.; Ernster, L.: Respiratory activity in isolated rat brain mitochondria following in vitro exposure to oxygen radicals. J. cerebr. Blood Flow Metab. *3:* 207–214 (1983).

8   Goldstein, S.; Czapski, G.: Mannitol as an .OH scavanger in aqueous solutions and in biological systems. Int. J. Radiat. Biol. *46:* 725–729 (1984).

9   Seligman, M. L.; Mitamura, J.; Shera, N.; Demopoulos, H. B.: Corticosteroid (methylprednisolone) modulation of photoperoxidation by ultraviolet light in liposomes. Photochem. Photobiol. *29:* 549–558 (1979).

10 Anderson, D. K.; Means, E. D.: Lipid peroxidation in spinal cord. $FeCl_2$ induction and protection with antioxidants. Neurochem. Path. *1:* 249–264 (1983).

11 Hunt, W. A.; Craig, C. R.: Alterations in cation levels and Na^+, K^+-ATPase activity in rat cerebral cortex during the development of cobalt induced epilepsy. J. Neurochem. *20:* 559–567 (1973).

12 Boehme, D. H.; Koseckis, R.; Carson, S.; Stern, F.; Marks, N.: Lipoperoxidation in human and rat brain tissue: developmental and regional studies. Brain Res. *136:* 11–21 (1977).

13 Seligman, M. L.; Flamm, E. S.; Goldstein, B. D.; Poser, R. G.; Demopoulos, H. B.; Ransohoff, J.: Spectrofluorescent detection of malonaldehyde as a measure of lipid free radical damage in response to ethanol potentiation of spinal cord trauma. Lipids *12:* 945–950 (1977).

14 Fourcans, B.; Jain, M. K.: Role of phospholipids in transport and enzymatic reaction; in Paoletti, Kritchevsky, Advances in lipid research, vol. 12, pp. 146–166 (Academic Press, New York 1974).

15 Goldman, S.; Alberts, R. W.: Sodium-potassium activated adenosine triphosphatase. IX. The role of phospholipids. J. biol. Chem. *248:* 867–874 (1973).

16 Rigoulet, M.; Geurin, B.; Cohadon, F.; Vandendreissche, M.: Unilateral brain injury in the rabbit; reversible and irreversible damage of the membranal ATPases. J. Neurochem. *32:* 535–541 (1979).

17 Wheeler, K. P.; Walker, J. A.; Barker, D. M: Lipid requirement of the membrane sodium-plus-potassium ion-dependent adenosine triphosphatase system. Biochem. J. *146:* 713–722 (1975).

18 Demopoulos, H. B.; Flamm, E. S.; Pietronigro, D. D.; Seligman, M. L.: The free radical pathology and the microcirculation in the major central nervous system disorders. Acta physiol scand., suppl. *492:* pp. 43–57 (1980).

19 Aust, S. D.; Svingen, B. A.: The role of iron in enzymatic lipid peroxidation; in Pryor, Free radicals in biology, vol. V, pp. 1–28 (Academic Press, New York 1982).

20 Fridovich, I.: Oxygen radicals, hydrogen peroxide and oxygen toxicity; in Pryor, Free radicals in biology, vol. I, pp. 239–277 (Academic Press, New York 1976).

21 Graf, E.; Mahoney, J. R.; Bryant, R. G.; Eaton, J. W.: Iron-catalyzed hydroxyl radical formation. Stringent requirement for free iron coordination site. J. biol. Chem. *259:* 3620–3624 (1984).

22 Witting, L. A.: Vitamin E and lipid antioxidants in free-radical initiated reactions; in Pryor, Free radicals in biology, vol. IV, pp. 295–39 (Academic Press, New York (1980).

23 Del Maestro, R. F.; Thaw, H. H.; Bjork, J.; Plaker, M.; Arfors, K. E.: Free radicals as mediators of tissue injury. Acta physiol. scand., suppl. *492:* pp. 43–57 (1980).

24 Freeman, B. A.; Crapo, J. D.: Biology of disease. Free radicals and tissue injury. Lab. Invest. *47:* 412–426 (1982).

25 Braughler, J. M.: Inhibition of synaptosomal GABA uptake by lipid peroxidation and Ca^{2+}: antioxidant effects of glucocorticoids (Abstract). Neurosci. Abstr. *10:* 942 (1984).

26 Hall, E. D.: High dose glucocorticoid treatment improves neurological recovery in head-injured mice. J. Neurosurg. (in press).

D. K. Anderson, PhD, Neurology Service (127), VA Medical Center,
3200 Vine Street, Cincinnati, OH 45220 (USA)

Novelli, Ursini (eds.), Oxygen Free Radicals in Shock. Int. Workshop,
Florence 1985, pp. 231–235 (Karger, Basel 1986)

Oxygen Free Radical Damage in the Critical Patient

O. Ortolani, A. Conti, R. Cuocolo

Università degli Studi di Napoli, II Facoltà di Medicina e Chirurgia,
Istituto di Anestesiologia e Rianimazione, Napoli, Italy

Introduction

Free radical mechanisms play fundamental roles in a large number
of physiological reactions, such as mitochondrial and microsomal ener-
getic pathways, oxygen transportation by hemoglobin, phagocytosis, etc.
Therefore, their existence is essential for the living matter. But when free
radicals are overproduced in a site where there is a poor scavenging activity, they may affect metabolic pathways other than the normal ones,
causing severe damages which can be summarized as follows: (1) biomembrane unsaturated fatty acid peroxidation, responsible for the alterations of membrane permeability, of the cell volume, of Na/K ratio, reduction of sarcoplasmic Ca^{++} uptake in the muscle and of membrane potentials in the liver and muscle, mitochondrial swelling, erythrocyte membrane weakening and stiffening; (2) protein denaturation, depolymeriza-
tion of glycosaminoglycans (hyaluronic acid) in and within the cell membranes and collagen denaturation with increased vascular permeabil-
ization; (3) lysosomal membrane injury and release of lytic enzymes; (4)
inhibition of physiological anti-enzymes (antitrypsin, etc.); (5) activation
of arachidonic acid cascade due to phospholipase enhancement by OH·
radicals; (6) arachidonic acid peroxidation to endoperoxides, thromboxanes and leukotrienes which increase vascular permeability and induce
vasoconstriction and platelet aggregation.

If one considers in detail the origin of oxygen free radicals in shock
and the role played by these radicals in some pathologic conditions, such
as disseminated intravascular coagulation (DIC) and adult respiratory

distress syndrome (ARDS), the results of the experiments performed on animals can provide very useful models: (1) Rats affected by endotoxic shock show reduced SOD levels [*Ogawa* et al., 1982] due to hypoxia and acidosis which hamper catalase and glutathione peroxidase activity and favor O_2^- and OH^- formation. (2) Hypoxic conditions increase the monoxigenase/cytochromooxidase ratio in animal mitochondria [*Edwards*, 1984], that means O_2^- hyperproduction. Complete ischemia with anoxia is less harmful to tissues than hypoperfusion with hypoxia, because free radicals cannot be formed without oxygen. (3) Some kinds of shock (endotoxic shock) are linked to overproduction of arachidonic acid vasoactive metabolites [*Demling* et al., 1981]. Furthermore, two products of arachidonic acid cascade, 12-*L*-hydroxi-eicosatetraenoic acid (HETE) and 12-*L*-hydroxi-5,8,10-eptadeca-trienoic acid (HHT), possess chemotactic activity for polymorphonucleate leukocytes [*Perez* et al., 1980]. (4) Enzymatic generation of these chemotaxins involves intermediate procoagulant peroxides which have also complement-activating properties [*Maier and Ulevitch*, 1981]. Prolonged survival may be obtained in endotoxic shock, by administering inhibitors of the arachidonic acid [*Halusha* et al., 1981]. (5) Hypoxia, shock and endotoxins induce the release of oxygen free radicals and H_2O_2 in a short circuit in which also the activated fraction C_5 of the complement is involved. Complement activation in critical patients may occur due to endotoxins, lysosomal enzymes, kinins, plasmin, hypoxia and acidosis. (6) Complement activation may be seen in lung embolization after hemodialysis or extracorporeal circulation. The leukoemboli produce free radicals which permeabilize the pulmonary endothelial wall and peroxidate the lipids of the alveolar membranes and of the surfactant. (7) Free radical scavenger mechanisms could prevent the mentioned complications. Because the lethal evolution of many shocks are frequently linked to a poor organic free radical-scavenging activity, an artificial supply of antioxidants should prove very effective in such cases [*Dillard* et al., 1977].

In the critical patient antioxidant vitamin deficit is a rule and the muscle catabolism, which always occurs, is associated with an increase of free iron. The iron may act as a positive catalyst in some free radical-generating reactions.

Both reduced natural scavenging and increased free radical production concur to determinate membrane destruction, lytic enzyme release, complement activation, granulocyte aggregation, stimulation of the hemocoagulative cascade.

Materials and Methods

In spite of a mess of experiments performed on animals, very few works deal with the role played by oxygen free radicals in the critical patient or with the effects of therapeutic administration of antioxidant drugs.

In the last 2 years we have tested over 200 intensive care patients whose clinical conditions were severely impaired after surgery, trauma, stroke, infections. The following parameters have been considered: blood gas values (Corning 175 analyzer); coagulative hyperactivation (fibrinopeptide A; Biochemia Elisa); C_{5a} production (granulocyte aggregation test) [*Jacob* et al., 1980]; plasma malonaldehyde concentration (thiobarbituric acid, TBA, test) [*Asakawa and Matsushita*, 1981]; ethane and pentane exhalation (gas-chromatographic method described elsewhere) [*Lawrence and Cohen*, 1984; *Ortolani* et al., in press].

The patients whose total bilirubin exceeded 1,2 mg/100 ml were excluded due to bilirubin interference with the TBA test. It is well recognized that the last three parameters are expression of the extent of a peroxidative process sustained by an excess of free radicals in the living organism.

Results and Discussion

The results obtained, compared with the values found in a group of 50 healthy volunteers, are shown in table I.

The statistical evaluation of these results showed an high incidence of complement activation, hypercoagulative status, and hyperproduction of free radicals (as measured through lipid peroxidative byproducts) in intensive care patients and a strong correlation between the related parameters.

It seemed of interest to assess the effects of antioxidant drugs on our critical patients. Three antioxidants were administered at the same time

Table I. Comparative valves of some parameters between normal subjects and intensive care patients

	Healthy volunteers (n = 50)	Intensive care patients (n = 200)	% increase	p
PO₂, mm Hg	89±16	61±29	−31	<0.005
FPA, ng/ml	1.5±0.7	9.1±7.8	500	<0.001
C_{5a}-aggregating index, %	22±4	43±15	105	<0.001
MDA-TBA reactants (OD)	0.030±0.008	0.070±0.040	103	<0.001
Pentane, nm/l	2±1.5	6.1±6.5	205	<0.001
Ethane, nm/l	0.8±0.5	2.8±2.5	205	<0.005

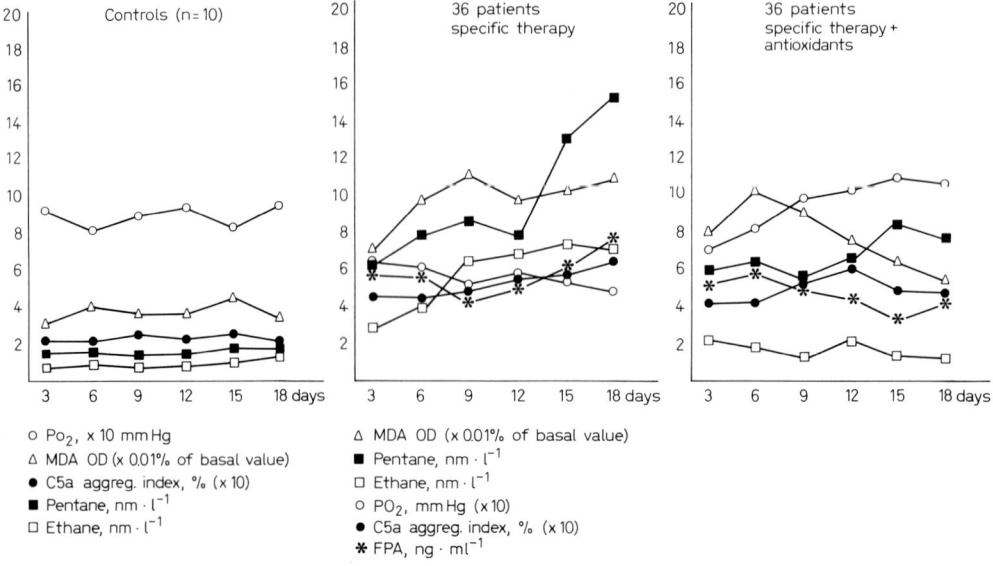

Fig. 1. Comparative parameters between a group of intensive care patients receiving only a specific therapy and a group supplemented with antioxidants.

in small doses in order to take advantage of the synergy of their actions. The vitamin E is the primary antioxidant and the resulting vitamin E radical reacts with vitamin C to regenerate vitamin E [*Packer* et al., 1978] while thiols potentiate their scavenging actions [*Copeland*, 1978].

A group of 72 intensive care subjects in very poor conditions was chosen. They showed high levels both of plasma malondialdehyde and of exhaled alkanes. These patients received a normal therapy for their specific pathologies and 36 of them received also vitamin E 30 mg/kg/die, vitamin C 50 mg/kg/die, and α-mercaptopropionylglycine 5 mg/kg/die.

The results obtained in these two randomized groups, and in a control group of 10 healthy volunteers receiving the same amount of antioxidants, are shown in figure 1.

The statistical evaluation performed by the Student's t test reveals a significant reduction in free radical production and free radical damages measured as lipid peroxidation, complement activation and coagulative unbalance in the group receiving antioxidants (p is always <0.01).

Some differences are visible between the patterns of ethane and pentane. Ethane is completely excreted whereas pentane is metabolized in the liver. Liver damage is frequently developed by critical patients, so that pentane metabolism is often reduced. This event may account for a delayed, diphasic increase of the peak of pentane as compared with the peak of ethane.

Although the results obtained agree with the hypothesized protective role of antioxidants against free radical damage in these patients, much more work is needed in order to elucidate to which extent free radical scavengers in human therapy may prove useful in avoiding the fatal evolution of some clinical pictures.

References

Asakawa, T.; Matsushita, S.: Coloring conditions of thiobarbituric acid test for detecting lipid hydroperoxides. Lipids *15:* 137–140 (1979).

Copeland, E. S.: Mechanisms of radioprotection. A review. Photochem. Photobiol. *28:* 839–844 (1978).

Demling, R.; Smith, M.; Gunther, R.; Flynn, J. T.; Gee, M.: Pulmonary injury and prostaglandin production during endotoxemia in conscious sheep. Am. J. Physiol. *240:* H348–H353 (1981).

Dillard, C. J.; Dumelin, E. E.; Tappel, A. L.: Effect of dietary vitamin E on expiration of ethane and pentane by the rat. Lipids *12:* 109 (1977).

Edwards, R. H. T.: Metabolismo energetico tissutale in corso di anossia e shock. Acta anestesiol. ital. *35:* 56–69 (1984).

Halusha, P. V.; Wise, W. C.; Cook, J. A.: Protective effect of aspirin in endotoxic shock. J. Pharmacol. exp. Ther. *218:* 464–469 (1981).

Jacob, H. S.; Craddock, P. R.; Hammerschmidt, D. E.; Moldow, C. F.: Complement-induced granulocyte aggregation. An unsuspected mechanism of disease. New Engl. J. Med. *302:* 789 (1980).

Lawrence, G. D.; Cohen, G.: Concentrating ethane from breath to monitor lipid peroxidation in vivo. Meth. Enzym. *105:* 305–311 (1984).

Maier, R. V.; Ulevitch, R. J.: The response of isolated rabbit macrophages to lipopolysaccharide (LPS). Circul. Shock *8:* 165–181 (1981).

Ogawa, R.; Morita, T.; Kunimoto, F.; Fujita, T.: Changes in hepatic lipoperoxide concentration in endotoxemic rats. Circul. Shock *9:* 369 (1982).

Ortolani, O.; Conti, A.; Biasiucci, M.; Mazzarella, L.; Imperatore, R.: Free radical lipid peroxidation through expired ethane and pentane. An improved method. Boll. Soc. ital. Biol. sper. (in press).

Packer, J. E.; Slater, T. F.; Wilson, R. L.: Direct observation of a free radical interaction between vitamin E and vitamin C. Nature, Lond. *278:* 737–738 (1979).

Perez, H. D.; Weksler, B. B.; Goldstein, I. M.: Generation of a chemotactic lipid from arachidonic acid by exposure to a superoxide generating system. Inflammation *4:* 313–316 (1980).

O. Ortolani, MD, Via Carducci 42, I-80121 Napoli (Italy)

Novelli, Ursini (eds.), Oxygen Free Radicals in Shock. Int. Workshop,
Florence 1985, pp. 236–240 (Karger, Basel 1986)

Physiopathological Patterns and Evolutive Stages of Sepsis

*Ivo Giovannini, Giuseppe Boldrini, Carlo Chiarla, Marco Castagneto,
Giancarlo Castiglioni*

Centro di Studio per la Fisiopatologia dello Shock, CNR, Istituto di
Clinica Chirurgica, Catholic University, Rome, Italy

Data from more than 400 physiological measurements performed in
95 septic patients have been thoroughly analyzed and confronted with
data from nonseptic trauma patients, and have provided an extensive assessment of the simultaneous and sequential cardiorespiratory and metabolic alterations associated with sepsis and septic shock.

The measurements were based on the determination of respiratory,
hemodynamic and metabolic parameters by the expired, arterial and central venous gas tensions, and by complementary data [10].

The results have allowed to assess and re-define the systemic septic
response as a generalized process with simultaneously occurring multiple
organ dysfunctions, and with a tendentially evolutive nature of the alterations.

A first stage was characterized by evidences of a respiratory malfunction with abnormally high pulmonary shunt and wasted ventilation,
and with an aberrant relationship of shunt and wasted ventilation to pulmonary blood flow; these changes appeared to be interactively mediated
by a microanatomical damage to the lungs and by a functional, hemody-
namically mediated, ventilation/perfusion maldistribution [6, 9, 18].
There were concomitant evidences of cardiac malfunction with multiple
signs of myocardial sufference and distress (in spite of a hyperdynamic
cardiovascular syndrome which was prevalently compensatory for the increased energy expenditure and for the peripheral metabolic failure and
O_2 extraction impairment), or transient ischemia, with relatively frequent
evolution into overt myocardial depression [7, 9, 11]. There were additional evidences of hepatic dysfunction; this resulted, through the abnor-

mal metabolism of vasoactive mediators, in a loss of peripheral vascular tone and in an abnormal relationship of vascular tone to O_2 consumption and cardiac output [12]. From the metabolic point of view, this septic stage was prevalently characterized by hypermetabolism and increased O_2 consumption, protein hypercatabolism, evidence of increased substrate (preferentially fat) utilization [3, 8]. The series of physiopathological events observed at this stage tended to remain at a subclinical level, becoming clearly detectable only on the basis of the performed measurements (pulmonary shunt, dead space, cardiac index and work, peripheral vascular tone, O_2 extraction, etc.); there were associated, however, other signs of multiple organ dysfunction.

A second stage was characterized by a worsening of the respiratory abnormalities, by multiple evidences of impaired oxidative metabolism with a relatively low O_2 consumption and metabolic acidosis, by a more critical dependency of metabolic compensation on cardiovascular hyperdynamism, with a high rate of, and more severe consequences of, episodes of cardiac failure; there was evidence of more severe hepatic malfunction [3, 8].

A third stage was characterized by clinically evident and generalized multiple organ involvement and failure, and a rapid evolution into death.

The observed changes tended to take place simultaneously, were more severe from the beginning in nonsurviving septics, became progressively more refractory to treatment. There was evidence of concomitantly progressive malfunction of other organs and systems. Multiple organ failure, with a variable prevalence of the pulmonary component, accounted for 90% of deaths; myocardial depression prevaled as a cause of death in 8% of cases.

The simultaneously occurring involvement of multiple organs and functions, with a similarly evolutive nature of the lesional processes which has been found to complicate the persistence of systemic sepsis, calls for a similar generalized mechanism of organ damage and deterioration; this is also supported by multiple pathological findings of evolutive microanatomical disruption at cellular and subcellular level in different organs in sepsis. These considerations and multiple circumstancial and specific evidences, separately provided by different authors, of the effect of oxygen free radicals in determining several of the observed phenomena, support the notion that these substances may have a central role in the evolutive multisystemic aberration of sepsis. The pathological changes of sepsis, ranging from light microanatomical disruption with increased cell permeability to more severe endothelial or tissue injury, resemble often those attributed to the effect of oxygen free radicals. Besides, many circumstances in which radical generation may be increased and scavenging protection decreased (phagocyte and leukocyte activation, endotoxemia, acidosis, hypoxemia, etc.) [2, 4, 15, 16, 22] frequently occur in sepsis.

The role of oxygen free radicals on endothelial damage and pulmonary function changes has already been extensively implied, by addressing not only the radical-mediated microanatomical changes and the increased capillary permeability, but also the prostaglandin production and the pulmonary hypertension and hypoxemia seen in septic patients [1, 20, 21, 23].

The role of free radicals on myocardial injury has been confined more to the ischemia-reperfusion mechanism of damage [5]; nevertheless, the possibility that transient regional hypoxia may affect the myocardium in sepsis and septic shock, the established myocardial depressant effect of free radicals, and the fact that the free radicals mediate some of the postulated mechanisms of impairment in cardiac performance in sepsis [11, 14, 19], may subtend the recognition of a role of these substances also in the subclinical or overt loss of cardiac competence that characterize the disease.

In the liver, lipid peroxidation may occur, it being promoted by the factors associated with the persistence of systemic sepsis; endotoxin infusion has been found to cause experimentally hepatic liver peroxidation [17].

Particularly interesting is finally the finding that a multiple organ failure sequence very similar to that observable in clinical sepsis could be experimentally induced in rats by intraperitoneal injection of zymosan (a method inducing radical-mediated reactions) [13].

By combining the information on the physiopathological events of sepsis obtained from patient measurements, and the considerations in this text, it is possible to postulate that an increased action of the oxygen free radicals at different levels may be a major common element in the development of the septic evolutive multisystemic impairment. This could also be supported by additional evidences not mentioned in the text. The hypothesis is not just simply speculative, and motivates the need for directly addressing this topic in clinical investigations. The practical implications also relate to the possibility of devising new therapeutic modalities.

References

1 Bertrand, Y.: Oxygen-free radicals and lipid peroxidation in adult respiratory distress syndrome. Intensive Care Med. *11:* 56–60 (1985).

2 Bielsky, B. H. J.; Allen, A. O.: Mechanism of disproportionation of superoxide radicals. J. phys. Chem. *81:* 1048–1052 (1977).

3 Castagneto, M.; Giovannini, I.; Boldrini, G.; Nanni, G.; Pittiruti, M.; Sganga, G.; Castiglioni, G. C.: Cardiorespiratory and metabolic adequacy and their relationship to survival in sepsis. Circulatory Shock *11:* 112–130 (1983).

4 Demopoulos, H. B.; Flamm, E. S.; Pietronigro, D. D.; Seligman, M. L.: Free radical pathology and the microcirculation in the major central nervous system disorders. Acta physiol. scand., suppl. 492, pp. 91–119 (1980).

5 Gardner, T. J.; Stewart, J. R.; Casale, A. S.; Downey, J. M.; Chambers, D. E.: Reduction of myocardial ischemic injury with oxygen-derived free radical scavengers. Surgery, St Louis *94:* 423–427 (1983).

6 Giovannini, I.; Boldrini, G.; Pittiruti, M.; Nanni, G.; Castagneto, M.; Sganga, G.; Tramutola, G.; Ronconi, P.: Early respiratory changes in sepsis. Am. Rev. resp. Dis. *124:* suppl., p. 66 (1981).

7 Giovannini, I.; Boldrini, G.; Sganga, G.; Tramutola, G.; Castagneto, M.: The determinants of myocardial distress in sepsis and septic shock. Crit. Care Med. *11: 230* (1983).

8 Giovannini, I.; Boldrini, G.; Castagneto, M.; Sganga, G.; Nanni, G.; Pittiruti, M.; Castiglioni, G. C.: Respiratory quotient and patterns of substrate utilization in human sepsis and trauma. J. Parent. Ent. Nutr. *7:* 226–230 (1983).

9 Giovannini, I.; Boldrini, G.; Castagneto, M.; Sganga, G.; Tazza, L.; Chiarla, C.; Castiglioni, G. C.: Cardiorespiratory patterns in sepsis, septic shock and surgical trauma. Acta med. rom. *22:* 257–263 (1984).

10 Giovannini, I.; Boldrini, G.; Castagneto, M.; Chiarla, C.; De Gaetano, A. M.; Castiglioni, G. C.: Reference sources and computational data for cardiorespiratory monitoring by mass spectrometer in critically ill patients. Spectros. Int. J. *3:* 401–407 (1984).

11 Giovannini, I.; Boldrini, G.; Chiarla, C.; Tramutola, G.; Castagneto, M.; Castiglioni, G. C.: Myocardial overperformance and distress in sepsis (submitted).

12 Giovannini, I.; Boldrini, G.; Chiarla, C.; Castagneto, M.; Tramutola, G.; Castiglioni, G. C.: Early vascular tone changes in sepsis (submitted).

13 Goris, R. J. A.; Nuytinck, J. K. S.; Boekholtz, W. K. F.; v. Bebber, I. P. T.; Schillings, P. H. M.: Oxygen free-radicals and sepsis with multiple organ failure. An experimental model in the rat (these proceedings).

14 Manson, N. H.; Deardorff, M. B.; Eaton, L. R.; Hess, M. L.: Possible role of leukocyte-derived oxygen free radicals in the myocardial failure of sepsis; in Novelli, Ursini, Oxygen free radicals in shock. Int. Workshop, Florence 1985, pp. 169–174 (Karger, Basel 1986).

15 Novelli, G. P.; De Gaudio, A. R.: Oxygen free radicals in shock states; in Lewis, Haglund, European Shock Soc. Meet., Malmö 1983. Shock Research, pp. 31–41 (Elsevier, Amsterdam 1983).

16 Ogawa, R.; Morita, T.; Kunimoto, F.; Fujita, T.: Changes in hepatic lipoperoxide concentration in endotoxemic rats. Circulatory Shock *9:* 369–374 (1982).

17 Sakaguchi, S.; Kanda, N.; Hsu, C.: Lipid peroxide formation and membrane damage in endotoxin-poisoned mice. Microbiol. Immunol. *25:* 229–244 (1981).

18 Siegel, J. H.; Giovannini, I.; Coleman, B.: Ventilation: perfusion maldistribution secondary to the hyperdynamic cardiovascular state as the major cause of increased pulmonary shunting in human sepsis. J. Trauma *19:* 432–460 (1979).

19 Singal, P. K.; Kapur, N.; Dhillon, K. S.; Beamish, R. E.; Dhalla, N. S.: Role of free radicals in catecholamine-induced cardiomyopathy. Can. J. Physiol. Pharmacol. *60:* 1390–1397 (1982).

20 Slater, T. F.: Free-radical mechanism in tissue injury. Biochem. J. *222:* 1–15 (1984).

21 Taylor, A. E.; Martin, D.; Parker, J. C.: The effects of oxygen radicals on pulmonary edema formation. Surgery, St Louis *94:* 433–438 (1983).

22 Ward, P. A.: Role of toxic oxygen products from phagocytic cells in tissue injury. Adv. Shock Res. *10:* 27–34 (1983).

23 Wong, C.; Flynn, J.; Demling, R. H.: Role of oxygen radicals in endotoxin-induced lung injury. Archs Surg., Chicago *119:* 77–80 (1984).

Dr. Ivo Giovannini, Via Alessandro VII, 45, I-00167 Rome (Italy)

Novelli, Ursini (eds.), Oxygen Free Radicals in Shock. Int. Workshop,
Florence 1985, pp. 241–244 (Karger, Basel 1986)

Superoxide Anion Scavenger Activity of New Imidazole Derivatives

*P. Cremonesi[a], D. Strada[a], A. Banfi[a], G. Sportoletti[a], A. Giotti[b],
S. Brunelleschi[b], R. Fantozzi[b]*

[a] Research Group Italfarmaco, Milan; [b] Department of Pharmacology,
University of Florence, Italy

Superoxide dismutase has an imidazole ring as histidine at the active site. Histamine (10^{-8} to $10^{-4} M$) inhibits superoxide anion (O_2^-) production from human neutrophils stimulated by the chemotactic peptide N-formylmethionyl-leucyl-phenylalanine (FMLP) [1]. Complexes of imidazole with some metals (e.g. copper, iron) can affect oxy-radical reactions by direct dismutation. To verify whether newly synthesized imidazole derivatives, with the general formula

R = Alkyl, alkylaryl, acyl

can interfere with oxy-radical production from cellular sources or can exert scavenging effects by themselves or by forming metal complexes, we performed a series of experiments by using three different models of generating oxy-radicals: a chemical model (pyrogallol autoxidation); an enzymatic model (xanthine/xanthine oxidase); a cell model (isolated human neutrophils and human whole blood).

Materials and Methods

Chemical Method. Initial rate of pyrogallol ($100 \mu M$) autoxidation in phosphate buffer ($5 \text{ m} M$, pH 8.9) solution saturated with oxygen was determined spectrophotometrically at 420 nm in the absence (control) or in presence of metals ($4 \mu M$) and/or imidazole derivatives ($40 \mu M$). The data are expressed as percentage of control.

Enzymatic Method. Enzymatic generation of O_2^- was obtained by the xanthine ($90 \,\mu M$)/xanthine oxidase ($2 \,mU/ml$) system. Initial rate of cytochrome c ($8 \,\mu M$) reduction was followed spectrophotometrically at 550 nm in the absence (control) or in presence of $1-4 \,\mu M$ of Cu^{2+} (as $CuSO_4$), with or without imidazole derivatives ($1 \,mM$). The data are expressed as percentage of control.

Oxy-Radical Production from Human Neutrophils. Human neutrophils were isolated from healthy volunteers by standard techniques of dextran sedimentation, Ficoll-Paque (Pharmacia) gradient centrifugation and hypotonic lysis of erythrocytes, as described by *Fantozzi* et al. [1]. The cells were suspended in a buffered salt solution (138 mM NaCl, $2.7 \,mM$ KCl, $8.1 \,mM$ Na_2HPO_4, $1.5 \,mM$ KH_2PO_4, $1 \,mM$ $MgCl_2$, $1 \,mM$ $CaCl_2$, pH 7.4) supplemented with $1 \,mg/ml$ glucose and treated for 5 min with cytochalasin B ($5\mu g/ml$: Aldrich) before exposure to the drugs. Cells were incubated with drugs for 5 min before being challanged by $10^{-7} \,M$ FMLP (Serva) or by $10^{-5} \,M$ A23187 (Calbiochem) O_2^- production was continuously monitored spectrophotometrically for 5 min after the addition of the stimulus, by measuring superoxide dismutase (Boehringer Mannheim) – inhibitable cytochrome c (Boehringer Mannheim) reduction [2]. In some experiments drug effects were compared by using human whole blood as a source of oxy-radicals, according to the method described by *Bellavite* et al. [3].

Results were expressed as percentage of control. Cimetidine was obtained by Smith Kline & French.

The original imidazole derivatives used throughout the study (ITF 13, ITF 40, ITF 43) were synthesized by Italfarmaco SpA (Milan, Italy). Human serum albumin (HSA) was obtained by Behringwerke.

Results and Discussion

The results obtained by pyrogallol autoxidation show a 3-fold increase of oxy-radical production in the presence of Mn^{2+}, while oxy-radical generation was not significantly modified in comparison with control when the experiments were performed in the presence of Cd^{2+}, Fe^{3+}, Co^{2+}, Zn^{2+} and Cu^{2+} (data not shown). None of the imidazole derivatives affected by itself pyrogallol autoxidation. However, when they were tested along with Cu^{2+}, a 20–30% inhibition of the reaction was observed with each drug. By using xanthine/xanthine oxidase system, a significant decrease of oxy-radical generation was recorded in the presence of ITF 40 only (fig. 1). The discrepancy between the data with pyrogallol autoxidation (no effect of ITF 40) and those with the xanthine/xanthine oxidase system suggests that ITF 40 may act as an inhibitor of xanthine oxidase.

The dismutating effects of copper were slightly increased by the addition of ITF 13 and ITF 43, whereas they were not modified by imidazole and histamine (fig. 1). When the experiments were repeated in the

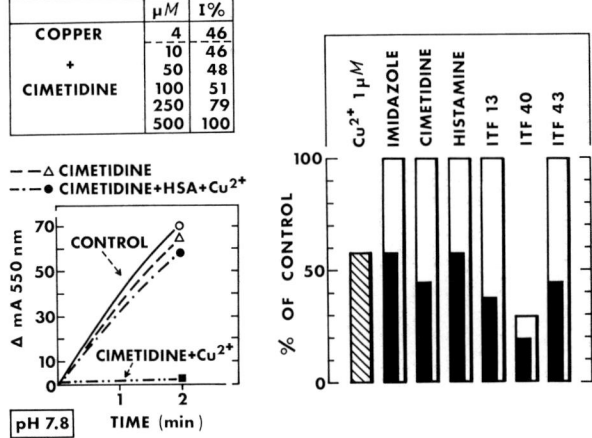

Fig. 1. Effect of Cu^{2+} and Cu-imidazole derivatives on oxy-radical production afforded by the xanthine/xanthine oxidase system. Blanc colums = drugs in the absence of Cu^{2+}; black colums = Cu-drug complexes. The panel on the left shows the reaction kinetics of 500 μM cimetidine and 4 μM Cu^{2+} with O_2^- as measured at 550 nm. I% = percent inhibition. n = 7.

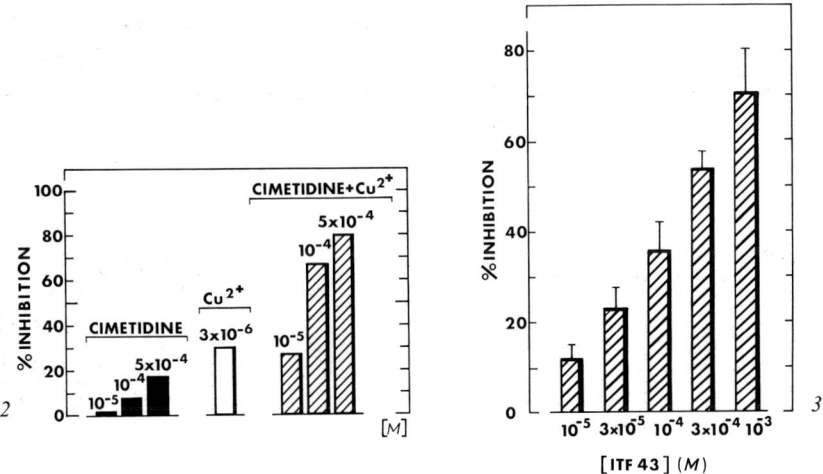

Fig. 2. Effect of Cu^{2+} and cimetidine on FMLP-evoked O_2^- production by isolated human neutrophils (n = 3). Means of values that varied $\leq 5\%$.

Fig. 3. Effect of ITF 43 on FMLP-evoked O_2^- production by isolated human neutrophils (n = 5). Means $\pm$ SEM.

presence of HSA (1 mg/ml) the scavenging effects of Cu^{2+} and of Cu-complexes were almost completely abolished for most of the drugs tested, thus indicating that HSA strongly competes with the complexes for the metal (fig. 1). However, the interaction between histamine and copper was not affected by HSA. Cimetidine, which is an imidazole derivative, strongly complexes Cu^{2+} as displayed by the difference spectra in the interval 360–320 nm (data not shown). Cu-cimetidine complexes exert superoxide dismutase (SOD)-like effects higher than those evoked by Cu^{2+} alone (fig. 1). The dismutating activity is lost in the presence of HSA.

When human neutrophils were exposed to 3 μM Cu^{2+} along with increasing concentrations of cimetidine, Cu^{2+} scavenging effects were greatly enhanced, being the drug by itself devoid of a significant inhibitory activity on FMLP-evoked O_2^- generation from neutrophils (fig. 2). ITF 43 dose-dependently inhibited O_2^- generation from isolated neutrophils stimulated by $10^{-7} M$ FMLP, with an inhibition of 70–80% at 1 mM (fig. 3). Similar inhibitory effects were given by the drug when O_2^- generation was evoked by samples of human whole blood challenged with $10^{-7} M$ FMLP or $10^{-5} M$ A23187. These results indicate that two different imidazole derivatives, cimetidine and ITF 43, affect oxy-radical production from neutrophils with different mechanisms. Cimetidine interacts with Cu^{2+} and displays a SOD-like activity. ITF 43 appears to inhibit directly the oxy-radical generation from the phagocytes.

References

1 Fantozzi, R.; Brunelleschi, S.; Giuliattini, L.; Blandina, P.; Masini, E.; Cavallo, G.; Mannaioni, P. F.: Mast cell and neutrophil interactions: a role for superoxide anion and histamine. Agents Actions *16:* 260–264 (1985).

2 Babior, B. M.; Kipnes, R. S.; Curnutte, J. T.: Biological defense mechanisms. The production by leukocytes of superoxide, a potential bactericidal agent. J. clin. Invest. *52:* 741–744 (1973).

3 Bellavite, P.; Dri, P.; Della Bianca, V.; Serra, M. C.: The measurement of superoxide anion production by granulocytes in whole blood. A clinical test for the evaluation of phagocyte function and serum opsonic capacity. Eur. J. clin. Invest. *13:* 363–368 (1983).

Prof. Roberto Fantozzi, MD, Department of Pharmacology, University of Florence, Viale G. B. Morgagni 65, I-50134 Firenze (Italy)

Subject Index